PHOTOVOLTAIC POWER GENERATION

PHOTOVOLTAIC POWER GENERATION

David L. Pulfrey, Ph.D.

The University of British Columbia

 VAN NOSTRAND REINHOLD COMPANY

NEW YORK   CINCINNATI   ATLANTA   DALLAS   SAN FRANCISCO
LONDON   TORONTO   MELBOURNE

Van Nostrand Reinhold Company Regional Offices:
New York   Cincinnati   Atlanta   Dallas   San Francisco

Van Nostrand Reinhold Company International Offices:
London   Toronto   Melbourne

Manufactured in the United States of America

Published by Van Nostrand Reinhold Company
135 West 50th Street, New York, NY 10020
Published simultaneously in Canada by Van Nostrand Reinhold Ltd.

15 14 13 12 11 10 9 8 7 6 5 4 3 2 1

Library of Congress Cataloging in Publication Data

Pulfrey, David L.
  Photovoltaic power generation.

  Includes bibliographical references and
index.
  1. Photovoltaic power generation. I. Title.
TK2960.P84   621.47'5   77-18220
ISBN 0-442-26640-5

To my children:

Simon,
Timothy,
Louisa.

Preface

It is now widely appreciated that the fuels and resources presently used in generating electricity may not be either sufficient or suitable to keep pace with the ever increasing world demand for electrical energy. The prospects for meeting this demand and averting crises in supply would be improved if alternative energy sources were to be developed. The Sun is one such source and conversion of sunlight directly into electricity via the photovoltaic effect is one method of generating solar electricity that has already received considerable attention. Photovoltaics has been successfully exploited both in space and in small-scale applications on Earth. But what is the likelihood of photovoltaics being used to generate electricity in amounts significant enough to affect the world's electrical energy picture? This book attempts to answer this question by giving a perspective to the possibilities and problems of photovoltaic energy conversion so that the factors governing its use in large-scale terrestrial electric power systems can be assessed. These factors are identified in the book and found to be of a technical, economic or institutional nature. Each of these aspects is examined in depth.

On the techno-economic side, the required components of a photovoltaic power system are presented and each item is discussed from the standpoint of present knowledge or performance, forseeable improvements and costs, required improvements for cost-effective power generation. The emphasis is on photovoltaic power systems appropriate to electricity supply at the residential, industrial and central power station levels in developed countries. However brief descriptions are also given of the small-scale photovoltaic power systems presently in use and of the mammoth-scale systems that may be viable in the early 21st century. From the arguments and information presented in the book it becomes apparent that it is of paramount importance to realize either lower cost levels or higher conversion efficiencies for the solar/electric transducer element of the power system before the large-scale use of photovoltaics can be contemplated. Accordingly the transducer, the solar cell, is treated in considerable detail. The basic theory and present performance of solar cells is reviewed and the current state-of-the-art in low cost solar cell fabrication and array production is critically discussed. The relevance of the concentrated sunlight approach to photovoltaic energy conversion is also discussed and the factors affecting the design and fabrication of solar cells suited to use with concentrators are examined.

The institutional factors governing the acceptance of photovoltaics as a large-scale electric power generation method refer to the laws and customs of the land and are examined in the book in the context of the established practices and interests of industrial, commercial and governmental bodies. Particular attention is paid to the role that existing electric utility companies might play in aiding the development of photovoltaics.

Encouragement of photovoltaic power generation now is postulated as being necessary to help ensure the orderly transition from the present fossil fuel-based electrical economy to a more durable one founded on solar and nuclear sources.

The scope of this book will be seen to be broad, reflecting the many factors involved in the development of photovoltaic power generation. Hopefully items of interest will be found by many of the correspondingly wide spectrum of people concerned about our future electrical prosperity, be they energy planners, administrators or educators, semiconductor research scientists and engineers, or technical laypersons. The sections of the book on photovoltaics in general (Chapter 1), on the constituents of large-scale photovoltaic power systems (Chapter 2), and on the economics of photovoltaic electricity (Chapter 6) need little more than an interest in solar electricity for their appreciation. The sections on the theory and development of solar cells (Chapters 3, 4 and 5) are more technical in nature and at least an elementary knowledge of solid state electronics would be desirable for their understanding.

I am grateful to the many people, too numerous to mention by name, who sent me details of their unpublished work and furnished me with data and diagrams from their publications. I would like to thank Drs. H. W. Dommel, P. J. Fenwick, M. M. Z. Kharadly and L. Young, all colleagues of mine at the University of British Columbia, for reading and constructively criticising various chapters of this book. My special thanks go to Ms. A. C. Hathorn who typed most of the manuscript in its various forms of draft and also provided very valuable editorial assistance. I thank also Ms. M. Hunter for typing some of the manuscript. Finally, I am indebted to my wife, Eileen, for her support, understanding and forbearance throughout the long period of writing.

D. L. PULFREY

LIST OF SYMBOLS

A^{**}	effective Richardson constant	$Am^{-2}\,°K^{-2}$
a	solar cell front surface total area	m^2
a_o	incident aperture area	m^2
B	semiconductor thickness	μm
C	solar concentration ratio	
D	diffusion coefficient	$m^2 sec^{-1}$
E	electric field	Vm^{-1}
E_g	semiconductor bandgap energy	eV
FF	fill factor	
F_M	photovoltaic system figure of merit	
fcr	fixed charge rate	
$G(\lambda,x)$	generation rate of minority carriers	$m^{-3} sec^{-1}$
h	Planck's constant	joule-sec
I_D	dark current	A
I_L	output current	A
I_m	current at maximum power point	A
I_o	saturation dark current	A
I_P	photogenerated current	A
I_{sc}	short-circuit current	A
i	composite mortgage interest and fuel-price escalation rate	
J_b	photocurrent density due to absorption in the base layer	Am^{-2}
J_b'	photocurrent density due to absorption of monochromatic light in the base layer	Am^{-2}
J_d	photocurrent density due to absorption in the depletion layer	Am^{-2}
J_d'	photocurrent density due to absorption of monochromatic light in the depletion layer	Am^{-2}
J_h	hole current density	Am^{-2}
J_i	photocurrent density due to absorption in the inversion layer	Am^{-2}
J_L	load current density	Am^{-2}
J_n	electron current density	Am^{-2}
J_{od}	injection-diffusion component of the saturation dark current density	Am^{-2}
J_{or}	recombination-generation component of the saturation dark current density	Am^{-2}
J_{os}	Schottky barrier saturation dark current density	Am^{-2}
J_P	photocurrent density	Am^{-2}
J_P'	photocurrent density due to absorption of monochromatic light	Am^{-2}
J_{Ps}	spectral photocurrent density	$Am^{-2}\mu m^{-1}$

J_s	photocurrent density due to absorption in the surface layer	Am^{-2}
J'_s	photocurrent density due to absorption of monochromatic light in the surface layer	Am^{-2}
J_{sc}	photocurrent density on short circuit	Am^{-2}
K_E, K_e	thermal conductance	$\text{kWm}^{-2}\,^{\circ}\text{C}^{-1}$
k	Boltzmann's constant	$\text{eV}\,^{\circ}\text{K}^{-1}$
L_h	hole minority carrier diffusion length	m
L_n	electron minority carrier diffusion length	m
$M(\lambda)$	photon flux	$\text{m}^{-2}\mu\text{m}^{-1}\text{sec}^{-1}$
m	Schottky barrier diode ideality factor	
m^*	carrier effective mass	kg
N_A	acceptor doping density	m^{-3}
N_D	donor doping density	m^{-3}
N_t	density of energy states available for tunnelling	$\text{eV}^{-1}\text{m}^{-3}$
n	excess electron density	m^{-3}
n_i	intrinsic carrier concentration	m^{-3}
n_R	refractive index of concentrating medium	
P_f	land utilization factor for an array	
P_i	solar power density	Wm^{-2}
p	excess hole density	m^{-3}
Q	array output fixed cost	$ \text{kWh}^{-1}$
q	charge on an electron	C
R_L	load resistance	ohm
R_s	solar cell series resistance	ohm
R_{sh}	solar cell shunt resistance	ohm
$R(\lambda)$	reflectance	
S_B	back surface recombination velocity	m sec^{-1}
S_F	front surface recombination velocity	m sec^{-1}
T	temperature	$^{\circ}\text{K}$
$T(\lambda)$	optical transmittance	
U	recombination rate of minority carriers	$\text{m}^{-3}\text{sec}^{-1}$
V_d	diffusion potential	V
V_j	voltage change at junction due to illumination or voltage bias	V
V_L	output voltage	V
V_m	voltage at maximum power point	V
V_{oc}	open-circuit photovoltage	V
V_{ox}	voltage developed across interfacial layer in an MIS cell	V
W	depletion layer depth	μm
X	module cost	$ \text{m}^{-2}$
x_i	inversion layer depth	μm
Y	capital costs (excluding module costs) associated with an array	$ \text{m}^{-2}$

Z	cost of land	$\$\ m^{-2}$
α	absorption coefficient	m^{-1}
γ	general factor appearing in the term $\exp(qV_j/\gamma kT)$ for the dark current	
δ	interfacial layer thickness	nm
ϵ	permittivity	Fm^{-1}
η	solar-electric conversion efficiency	
η_{coll}	collection efficiency	
η_{op}	total optical efficiency of concentrator system	
θ_c	acceptance angle	deg.
λ	wavelength	nm
λ_g	wavelength at the bandgap energy	nm
μ	mobility	$m^2\,V^{-1}sec^{-1}$
ξ	annual energy output of an array	$kWh\ m^{-2}\ yr^{-1}$
ρ_b	base layer resistivity	ohm cm
τ	minority carrier lifetime	μsec
ϕ_b	metal–semiconductor barrier height	eV
ϕ_m	metal work function	eV

Contents

Preface   vii
List of Symbols   ix

1. INTRODUCTION   1

2. ELEMENTS OF PHOTOVOLTAIC POWER SYSTEMS   9

     2.1 Insolation   12
     2.2 Photovoltaic Arrays   21
       2.2.1 Arrays for use in unconcentrated sunlight   22
       2.2.2 Arrays for use in concentrated sunlight   28
     2.3 Power Conditioning and Solar/nonsolar Power Plant Interconnection   37
     2.4 Energy Storage   42
     2.5 Present-day Photovoltaic Power Systems   51
     2.6 Geosynchronous Satellite Solar Power   56
     2.7 The Institutional Aspects of Photovoltaic Power Development   62

3. SOLAR CELLS: BASIC THEORY AND PRESENT PERFORMANCE   66

     3.1 Solar Cell Equivalent Circuit   69
     3.2 The Short Circuit Photocurrent   72
       3.2.1 Homojunctions   74
       3.2.2 Heterojunctions   79
       3.2.3 Schottky barriers   85
       3.2.4 Measured values of short circuit photocurrent   91
     3.3 The Open Circuit Photovoltage   93
       3.3.1 Homojunctions   94
       3.3.2 Heterojunctions   98
       3.3.3 Schottky barriers   100
       3.3.4 Measured values of open circuit voltage   103
     3.4 The Fill Factor   104
       3.4.1 Calculated values of fill factor   104
       3.4.2 Measured values of fill factor   107
     3.5 The Efficiency   108
       3.5.1 Calculated values of efficiency   108
       3.5.2 Measured values of efficiency   112

4. SOLAR CELLS FOR UNCONCENTRATED SUNLIGHT SYSTEMS   **114**

4.1 Conversion Efficiency in Large Area Solar Cells   116
   4.1.1 Series and shunt resistance losses   116
   4.1.2 Minority carrier properties   118
   4.1.3 Semiconductor thickness   120
   4.1.4 Grain properties in polycrystalline semiconductors   121
4.2 Silicon Solar Cells   124
   4.2.1 Solar-grade silicon   124
   4.2.2 Silicon wafer preparation   126
   4.2.3 Silicon sheet preparation   129
   4.2.4 Silicon film preparation   133
   4.2.5 Device processing techniques   137
4.3 Cadmium Sulfide Solar Cells   141
   4.3.1 Cu_2S/CdS cells   142
   4.3.2 Other heterojunction cells using CdS   145
4.4 Other Semiconductor Possibilities   147
   4.4.1 Inorganic semiconductors   148
   4.4.2 Organic semiconductors   152
4.5 Energy Payback Time   153

5. SOLAR CELLS FOR CONCENTRATED SUNLIGHT SYSTEMS   **154**

5.1 Conversion Efficiency in Concentrator Solar Cells   156
   5.1.1 High illumination intensity effects   156
   5.1.2 High temperature effects   159
5.2 Gallium Arsenide Solar Cells   162
5.3 Silicon Solar Cells   164
   5.3.1 Single junction devices   164
   5.3.2 Multijunction devices   167

6. ECONOMIC ASSESSMENT OF PHOTOVOLTAIC POWER SYSTEMS   **171**

6.1 Array Output Fixed Costs   175
6.2 The Cost of Electricity from Photovoltaic Power Systems   179
   6.2.1 Residential systems   179
   6.2.2 Intermediate-level systems   186
   6.2.3 Central power station systems   188

7. CONCLUSIONS   **197**

REFERENCES   **200**

INDEX   **211**

PHOTOVOLTAIC POWER GENERATION

1.

Introduction

Many societies across the world in which we live have developed a large appetite for electrical energy. This appetite has been stimulated by the relative ease with which electricity can be generated, distributed, and utilized, and by the great variety of its applications. It is arguable whether the consumption of electricity should be allowed to grow unchecked, but the fact is that there is an ever-increasing demand for this synthetic energy form. Clearly, if this demand is to be met, then the world's electricity generating capacity will have to continue to grow.

Presently almost all electricity generation takes place at central power stations which utilize coal, oil, gas, water or fissile nuclear materials as the primary fuel source. There are problems facing the further development of generating methods based on any of these "conventional" fuels. The continued large-scale use of oil and gas in countries not blessed with indigenous reserves is particularly doubtful because supplies are expensive, rapidly diminishing, and politically controlled. Hydro-electric power generation is restricted to geographically suitable areas, and reserves of coal, although presently plentiful, are not renewable. The possible hazards of nuclear power have been much publicized, particularly those concerning the storage and military use of nuclear waste material. Nevertheless, to assist in maintaining electricity supply standards in many of our societies its seems likely that an increasing nuclear power presence, involving breeder and possibly fusion reactors, will have to be tolerated. However, it is clearly desirable that ways be sought to minimize the proliferation of nuclear power plants. To achieve this and also to aid in the management of existing fossil-fuel resources, it is essential that some part, and an increasing part, of future electrical energy research and development be concerned with so-called "nonconventional" methods of generation.

Fuel cells, magnetohydrodynamic systems, and devices based on thermoelectric, thermionic and solar-electric conversion are all potentially useful nonconventional electricity sources. Each of these sources has its advocates for further development, but none more so than solar energy which capitalizes, perhaps, on the deep-rooted association between man and sun to foster an image of bountiful power from a non-depletable, nonpolluting and benign source. The potential of solar-electric conversion is immense and current research seeking to realize it involves studies on bioconversion, the wind, photovoltaics, ocean currents, photoelectrochemistry and photothermodynamics.[1] Whilst all these methods can be designed to yield electricity as the end product, if so desired, it is only through the photovoltaic effect that sunlight can be converted directly into electricity. This feature of directness of conversion has been largely responsible for making photovoltaics the method chosen for supplying electricity to nearly all the major satellites and spacecraft that have been launched. Although it is most unlikely that this method, or indeed any method, will ever enjoy such a monopoly in terrestrial generation, the experience gained in space operations has given photovoltaics a sound theoretical and experimental base on which to found attempts to supply electricity to Earth. Considerable advances have been made in this direction and to describe this progress is one of the aims of this book.

The heart of any photovoltaic power system is the solar cell. It is the transducer that converts the sun's radiant energy directly into electricity, and is basically a semiconductor diode capable of developing a voltage of 0.5–1 V and a current density of 20–40 mA cm^{-2}, depending on the materials used and the sunlight conditions. The connection of solar cells in series and parallel and incorporation into a module provides a higher rated unit which can be interconnected with similar modules to comprise an array. In principle, array sizes at the thousand megawatt level are possible, being limited mainly by real estate considerations. The maximum efficiency for the conversion of sunlight to electricity via the photovoltaic effect is around 25% and for unconcentrated sunlight conditions on Earth the maximum solar power intensity is close to 1 kWm^{-2}. Under these conditions the maximum possible electrical output is 250 W for every square meter of transducer. Thus an appropriately positioned residential rooftop of area 80 m² could, in principle, realize a peak electrical power of 20 kW whilst, at the central power station level, e.g., a peak rating of 2500 MW, a transducer area of at least 10 km² would be required. Coverage of such areas with space-type solar cell arrays is presently prohibitive on account of their high cost (about $ 10,000 m⁻², which at a typical space-sunlight conversion efficiency of 11.5%, yields a price of around $70 per peak watt). To reduce this $/W figure is the principal aim of most of the present work in terrestrial photovoltaics. Substantial national programs have been initiated in various parts of the world in an attempt to make progress in this direction (see Table 1.1). Already the achievements are considerable and in the United States, for example, solar cell arrays for terrestrial applications can presently be procured at a cost of $15 per peak watt. This represents a halving in price over the period 1974–76, and the prospects of marketing arrays at $0.50 per peak watt by 1986 appear reasonable. This latter figure represents the goal set by the United States Energy Research and Development Administration (ERDA) to provide a focus for their photovoltaic programs and is linked to the annual production of 500 MW_{pk} of solar cell arrays by 1986. A further ERDA goal is the production of 50,000 MW_{pk} of solar cell arrays in the year 2000 at a market price of $0.1–0.3 per peak watt. This represents electricity in the 10–30 mills kWh^{-1} price range and the installed capacity ($\sim$20,000 MW_{av}) would be about 1–3% of the total U. S. generation capacity at that time (i.e., 5% of the present total capacity, which is equivalent to the present output from U. S. nuclear plants). Bringing down the cost of solar

Table 1.1 Funds Allocated to Photovoltaic Research at the National and International Levels.[a]

	$M	PERIOD
France	1.3	1977
West Germany	2.2	1976–80
Japan	1.0	1977
United States	65.0	1977
European Economic Community	9.0	1974–79

[a] Data collected from Refs. 2 and 3.

cell arrays thus has to be compatible with high volume production methods.

However, the mass production of cheap solar cell arrays is not the only means whereby the price ($/W) of photovoltaic power can be reduced. An alternative approach is to accept the fact that solar cells are an expensive item, and to seek to increase their output by illumination with concentrated sunlight. The solar collector system in this approach comprises the solar cell and reflective or refractive concentrating components, and possibly also facilities for tracking the sun and cooling the solar cells. If the concentrating materials can be made cheaper than the solar cells they replace, then the extra costs involved in tracking and cooling can perhaps be tolerated, thus leading to the possibility of photovoltaic electricity at a lower cost than is attainable with the unconcentrated sunlight approach. The prime requirement for solar cells operating in concentrated sunlight is one of high conversion efficiency, whereas in the unconcentrated sunlight approach it is the trade-off between efficiency and cost per unit area that is important. The high efficiency must be maintained under conditions of high illumination intensity and, quite probably, elevated temperatures.

Solar cells are obviously only able to produce electricity when the sun is shining on them and so, to maintain continuity of supply, it is necessary that the photovoltaic power system include either an energy storage unit or a tie-in with a nonsolar generating plant. Both of these latter items are likely to be present in photovoltaic systems designed to supply power to loads typical of modern residential, commercial and industrial consumers. Under these circumstances a displacement of conventional generating capacity and an improvement in system load factor are possible. These factors are obviously of great importance to existing electrical utilities, but photovoltaics may also seem attractive to the consumer because it could offer a measure of independence from the utility. An early resolution of such differences of interest is desirable for the acceptance of photovoltaics once it attains techno-economic viability.

Through the utilization of photovoltaic systems to provide power at the residential, commercial and industrial levels some impact of solar electricity on the generation mix currently employed by utilities can be expected. Although this time is perhaps 15–20 years away (even in the United States where research and development activity is high), there are areas where photovoltaics has already been found to be cost-competitive with other means of generating electricity. The wide variety of present applications listed in Table 1.2 have in common one or more of the following factors that render photovoltaic power attractive: the reliability of operation; the need for only infrequent maintenance; the lack of pollution; the absence of costs associated with expendable fuel and its transport to and storage at the site; the remoteness of the location; the portability; the modular nature of solar cell arrays; the ease of installation. In addition, the applications are characterized by the relatively small electrical size of the loads and the fact that adequate supply can be achieved either from the photovoltaic array alone or from a simple solar array-storage battery combination.

The applications listed in Table 1.2 attest to the sizable market now existing for terrestrial power units. Continued penetration into the general fields covered in Table 1.2 will necessitate increases in solar array production levels. This, in turn, will assist in bringing down array prices and in increasing public awareness of the capabilities of terrestrial photovoltaic power. It has been estimated that to reach the 1986 goal of arrays at $0.50 per peak watt by cost reduction based on manufacturing learning and experience alone, would necessitate a tripling of the solar cell market (and array production levels) each year for the next ten years.[13] For this to occur new markets would have to be identified and supplied. The installation of photovoltaic power systems in sunshine-rich underdeveloped countries could generate an enormous market. The advent of electric vehicles for urban transport would provide a large market for photovoltaic-powered charging systems, with the location of solar cell arrays on roofs over parking lots, carports and garages being envisioned. In the military sector the tactical, ancillary, mobile and remote applications of photovoltaic power systems make their use very attractive. In the United States the Department of Defense is already contributing significantly to the development of terrestrial photovoltaics and the decision to interface a photovoltaic power system with the electric utility network in Bermuda to power a 100 kW load at a U. S. Navy research laboratory augurs well for the future of photovoltaics.[7,14] Here experience with photovoltaic power systems

Table 1.2. Examples of Present Terrestrial Applications of Photovoltaic Power Units.

POWER SUPPLY APPLICATION	PEAK RATING, W	REF.
Warning lights:		
airport light beacon	39	4a
marine light beacon	90	4a
railroad signals		4e
highway barrier flashers	1.2	5
tall structure beacon		4c
lighthouse		6
Communications systems:		•
remote repeater stations for		
—microwaves	50	3
—radio	109	7
—TV	78	4a
remote communications station	3500	4c
mobile telephone communication station	2400	7
portable radio	50	5
emergency locator transmitter		4d
Water systems:		
pumps in desert regions	400	3
water purification	10800	7
Scientific instrumentation:		
telemetry—collection and transmission platforms for environmental, geological hydrological and seismic data		4b
anemometer	100	5
remote pollution detectors—H_2S		3
—noise	3	5
Industrial:		
remote machinery and processes, e.g., copper electrolysis installation	1500	4a
cathodic protection of underground pipeline	30	8
electric fence charger		9
domestic water meter	20	3
off-shore drilling platforms		3
forest fire lookout posts		9
Battery charging:		
boats, mobile homes and campers, golf carts	6–12	4d
construction site equipment		4d
Ni–Cd-powered military equipment	74	7

Recreational and educational:

educational TV	35	10
vacation home		
—lighting, TV		4c
—refrigerators	200	11
sailboats		
—lighting, ship-shore communication		4d
—automatic pilot	66	3
portable TV camera		12
camping lighting		4d
electronic watches, calculators		4b
recreational center sanitary facility	168	4b
Security systems:		
closed circuit TV surveillance	150	5
intrusion alarms	6	5

closer to the size and nature germane to future civilian utility use will be gained.

Eventually, on demonstration of reliable performance and on attainment of the production and economic goals described above, photovoltaic power systems will doubtless begin to appear in generation systems supplying power to residential, commercial and industrial loads. The information presented in the following pages attempts to place the problems and possibilities of photovoltaics in perspective, so that the factors governing its use in such large-scale power systems can be assessed.

The basic elements that would comprise a large-scale photovoltaic power system are discussed in the next chapter. At that stage the solar cell is considered only as a "black-box" component of an array. However, because it is of paramount importance to realize either lower cost levels or higher efficiencies for these units before the large-scale use of photovoltaics can be contemplated, it becomes necessary to consider the solar cell in some detail. This is done in Chapters 3–5. Firstly the basic theory and present performance of solar cells is reviewed, then the current state-of-the-art in low-cost solar cell fabrication and array production is discussed. For solar cells suited to use in concentrator systems it is the solar cell conversion efficiency rather than its cost that is the salient parameter; the factors affecting the design and fabrication of such cells are discussed in Chapter 5. Armed with this knowledge on photovoltaic systems' requirements and solar cell performance, it is possible to make some economic forecasts relevant to the incorporation of photovoltaics into large-scale electricity generating systems. Some predictions are presented in Chapter 6, so enabling some conclusions to be drawn about the role photovoltaics might play in future electrical energy scenarios.

2.
Elements of Photovoltaic Power Systems

Photovoltaic power units that are in use today generally comprise a relatively simple arrangement of solar cell array, blocking diode, storage battery (perhaps with overcharge protection) and a load. Figure 2.1 shows, as an example, the setup for powering a remote television station.[10] This arrangement is satisfactory for supplying small electrical loads that have a reasonably predictable demand. The nonuniformity of the local insolation is usually accounted for by simply sizing the power unit on the basis of likely worst-case conditions. There are many applications for which this procedure, and power units similar to that shown in Fig. 2.1, are suitable (see Table 1.2), and these will continue to provide a market for terrestrially based photovoltaic powerpacks. However, the renewable and nonpolluting nature of the photovoltaic energy source make its incorporation into larger electrical systems highly desirable. But the electrical demands of modern residential, commercial and industrial loads are large and variable, so much so that there is little likelihood of meeting them solely via units of photovoltaic arrays and battery stores. For photovoltaics to make any contribution to large-scale electricity generation, the photovoltaic powerpack needs to be extended into a system that includes solar cells and energy store and also takes into account the variability of the input energy source and the possibility of interconnection with a nonsolar generating plant. The elements that would have to be incorporated into such a terrestrially located photovoltaic power system are depicted in Fig. 2.2.

9

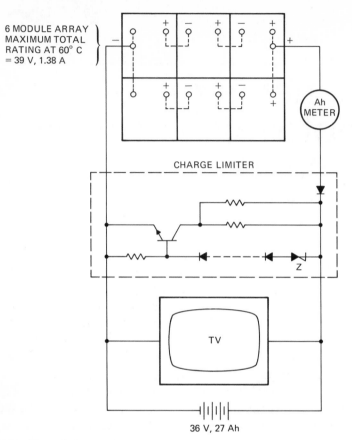

Fig. 2.1. 50 W photovoltaic power system for a remote television station. (Adapted from Polgar[10])

The insolation is dependent on the location and orientation of the solar cell array and provides an input energy to the system that has medium and long-term variations due to diurnal and seasonal factors respectively, as well as short-term variations due to local climatological conditions. The electrical output from the solar array is thus in the form of a variable dc power which needs some conditioning before it can be fed to the loads typical of the present-day consumer. The power conditioner is likely to contain not only an inverter for developing ac power, but also some form of input power-tracking device in order that the solar cell module can operate continually at its maxi-

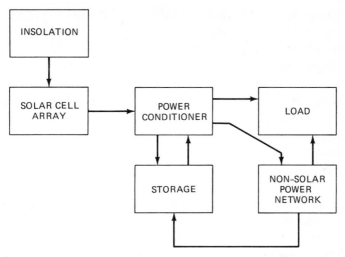

Fig. 2.2. Block diagram of the elements of a large-scale photovoltaic power system.

mum power point. In order not to waste any electrical power that is generated in excess of the immediate demand, and also to feed power to the load when the demand exceeds the immediate generation level, some form of energy storage is necessary. Finally under those circumstances where the photovoltaic power unit does not meet the demand all the time, some interconnection with a nonsolar electricity generating plant is necessary if the presently-accepted continuity of supply is to be maintained.* The possible interconnections shown in Fig. 2.2 allow for not only direct supply of nonsolar electricity to the load but also for supply to the storage system and transfer of excess solar electricity to the grid. The location of the store, the control over its dispatch and the nature of the connections to it are important considerations for the integration of the solar power plant into the electricity grid network. The eventual aim would be to have the photovoltaic power system bring about some displacement of conventional generating capacity.

---

*Present utility standards expect a power station outage not more than once every ten years, and even then the grid network is such that this outage is intended to be accommodated so as not to lead to a curtailment of power to the consumer.

The above items constitute a skeleton of the photovoltaic power systems expected to make some impact on the generation mix currently used for the supply of electricity on a large scale. The description of these elements forms the main subject matter of this chapter. Brief mention is also made of the salient features of the small-scale photovoltaic power systems currently in use, and of the mammoth-scale systems that might practicably deliver power to Earth from space sometime in the twenty-first century.

2.1 INSOLATION

The worldwide distribution of solar energy in terms of duration of sunshine is shown in Fig. 2.3. The bold contour lines enclose regions in which there is an average of about 8 or more hours of sunshine a day. With the exceptions of the Southern United States, South Africa and Australia, these regions comprise relatively underdeveloped Third World countries, and it is in these latter areas where considerable benefit could be gained immediately from the widespread utilization of solar energy, even via photovoltaic systems at their present small size. In these countries a 30–50 kW array in a village could go a long way toward improving the environment via lighting of homes and the pumping of water. Already France is reportedly supplying thousands of photovoltaic-powered water pump units to North Africa, and a French-supplied solar cell-powered TV network is being developed for educational purposes in Niger.[3, 10] This may border on technological imperialism but attempts are being made by the United Nations to incorporate similar high-technology solar schemes into more diversified solar energy demonstration projects in Asia, Africa and Latin America.[16]

The export potential of photovoltaic systems seems immense and could conceivably begin to be realized almost immediately. In the more distant future another form of solar energy exportation, namely in the form of either pressurized hydrogen from photovoltaic-powered electrolysis plants or microwave energy from photovoltaic electricity (beamed off power relay satellites in geosynchronous orbit),[17] could lead to a new balance of power amongst the energy producing countries. Australia with its vast area, abundant sunshine and meager population has a particularly favorable energy export potential. In the

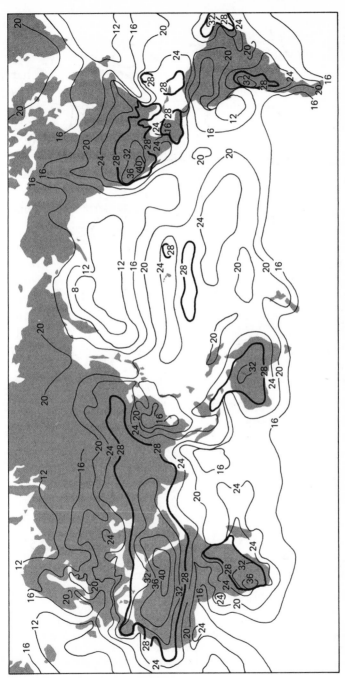

Fig. 2.3. Worldwide distribution of solar energy in terms of duration of sunshine. The contours on the figure refer to insolation in hundreds of hours per year. (After "Solar Energy" cited in Ref. 15; courtesy of Pergamon Press)

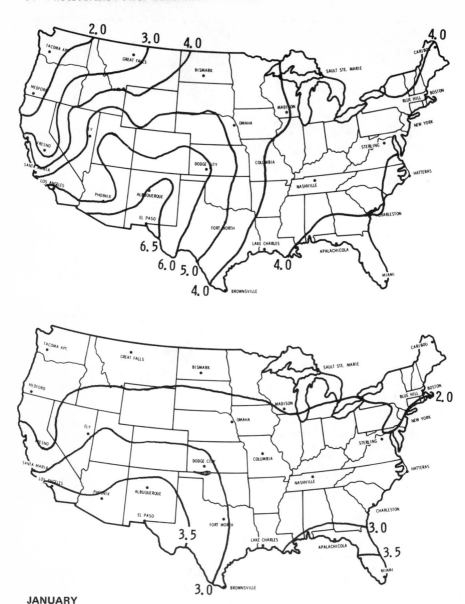

Fig. 2.4.   Mean daily solar radiation in the U.S. for January and June.   Top figures are for the direct component incident on a normal plate, bottom figures are for total radiation on a horizontal plate.   Contours are kWh m^{-2}.   (After Boes et al.;[18] courtesy of Amer. SES)

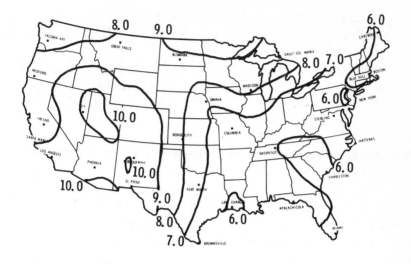

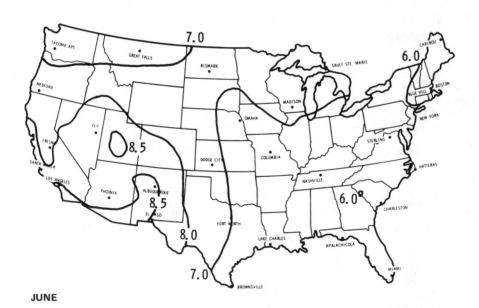

JUNE

Fig. 2.4 (*continued*)

United States where photovoltaic conversion technology is well-advanced, most attention has been given to assessing the role photovoltaics can play in providing domestic power. The south-western portion of the country is particularly well endowed with insolation as can be seen from the plots of Fig. 2.4. However the decrease in collected solar radiation at higher latitudes can be substantially reduced by orientating the solar collectors to the south and tilting them at an angle equal to the local latitude. The results of calculations for such conditions based on measurements of horizontally-collected data are shown in Table 2.1. By arranging for the solar collector to track the sun continually a further increase in insolation could be achieved, e.g., at Albuquerque, New Mexico (40° lat.) a fixed flat-plate collector tilted at 40° would receive about 25% less insolation per year than a polar mounted tracking flat-plate.[20] If collectors are designed to concentrate the sunlight then only the direct-normal component of the insolation is of importance, and so such installations would not be suitable in locations with frequent cloud cover. Phoenix, Arizona is not such a place since about 80% of its solar energy is in the form of direct radiation, and calculations for this location indicate that a two-axis tracking aperture would receive 20% more energy (all direct) than an optimally positioned fixed surface collecting both direct and diffuse radiation, see Fig. 2.5.[21]

Table 2.1. Insolation Data for Selected Locations in North America.[a]

| | | MEAN DAILY INSOLATION (kWhm^{-2}) | |
LOCATION	LATITUDE	HORIZONTAL	TILTED
Panama Canal Zone	8° 39′	4.90	4.96
Honolulu HI	21° 8′	5.94	6.27
Miami FL	25° 49′	5.27	5.78
Charleston SC	32° 54′	4.70	5.40
Phoenix AZ	33° 26′	6.05	6.98
Sayville NY	40° 46′	4.07	5.01
Cleveland[b] OH	41° 39′	3.93	4.77
Ottawa ONT	45° 27′	3.86	4.99
Seattle WA	47° 27′	3.50	4.39
Churchill MAN	58° 45′	3.22	5.19
Fairbanks AK	64° 49′	2.75	4.77

[a]Based on data from Ref. 19.
[b]Put-in-Bay weather station.

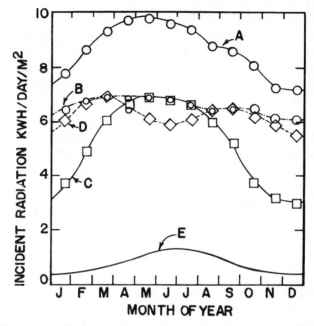

Fig. 2.5. Yearly distribution of direct solar radiation incident on surfaces of various orientations.   Data are for cloud-free conditions at Phoenix, Arizona. (After Backus[21] with permission).

Curve	Description	Energy Collected
A	Sun following	3170 kWh/yr/m^2
B	Non-sun following but with daily adjustment at noon	2400 kWh/yr/m^2
C	Horizontal surface (33° N. lat.)	1950 kWh/yr/m^2
D	Fixed area tilted at local lat.	2250 kWh/yr/m^2
E	Diffuse radiation on horizontal surface	400 kWh/yr/m^2

To evaluate the performance of photovoltaic converters it is necessary to know not only the amount of sunlight that can be collected but also its typical spectral distribution and how this is influenced by atmospheric conditions. Some information is given by classifying sunlight conditions on the basis of an air mass (AM) number, where the number refers to the secant of the angle between the positions of the solar beam at the zenith and at the time under consideration, measured at sea level. For example, when the sun is overhead the angle is $0°$ and conditions can be classed as AM1, and when the sun has moved by $60°$ from the zenith AM2 conditions prevail. The air mass number is therefore a measure of the thickness of the atmosphere between the observation point and the sun, and it is this particular interpretation of air mass that allows the classification AM0 to be used for sunlight outside the earth's atmosphere, taken for reference at the earth's mean distance from the sun. For situations on the earth's surface the description must be extended to account for the prevailing weather conditions and the absorption and scattering of light in the upper atmosphere. Scattering is due to air molecules and aerosols, such as dust and water droplets, and absorption is due principally to ozone and water vapor, with lesser contributions coming from oxygen and carbon dioxide. In addition to direct and diffuse components of radiation, terrestrial sunlight also has a component due to reradiation from buildings, vegetation, water etc. and subsequent backscattering from the atmosphere. However this latter radiation is usually of wavelengths in the range 6–60 μm, which are too long to be of importance in photovoltaics.

A detailed analysis of how many of the above factors affect the irradiance relevant to solar cells has been presented by Böer,[22] who proposed five sets of insolation conditions to cover the likely terrestrial, clear-weather, operating range. This scheme will likely prove of importance in future calibration work but at present most workers in the photovoltaics field quote results in terms of standard spectra, which for the extraterrestrial case are based on Thekaekara's data,[23] and for terrestrial conditions are often based on the data of Moon[24] or of the International Commission on Illumination.[25] The standard United States spectra are shown in Fig. 2.6.[26] The proposed European standard uses slightly different values for the atmospheric parameters leading to the AM1 spectrum given in Table 2.2.[25] There is

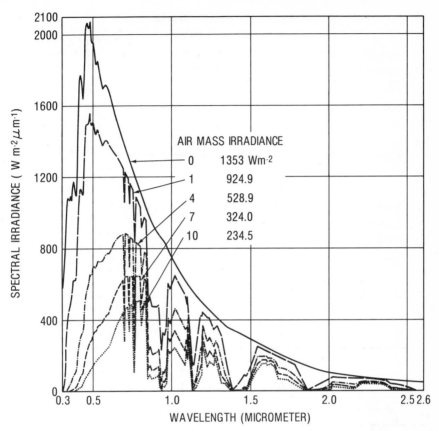

Fig. 2.6. Solar irradiance for different air mass values under U.S.A. standard atmospheric conditions; $H_2O = 20$ mm; ozone = 3.4 mm; turbidity factors—$\alpha = 1.3$, $\beta = 0.04$. (After Thomas et al.,[26] courtesy of Amer. SES)

clearly a need for an internationally approved standard for calibrating and testing solar cells, and the recommendation of a recent ERDA workshop that addressed this situation is to adopt a solar spectrum close to AM1.5 for this purpose.[26a] At the laboratory level a common procedure is to simulate terrestrial sunlight by the simple expedient of filtering the light from a 3200°K color temperature tungsten-halogen lamp through water and adjusting the intensity until a reference solar cell detector indicates a power density of 100 mWcm^{-2}.

Table 2.2. AM1 Irradiance.[a]

RANGE	WAVELENGTH (μm)	IRRADIANCE (W/m^2)		PERCENTAGES OF TOTAL RADIATION (%)	
0	< 0.28	0		0	
	0.28–0.32	5		0.5	
1	0.32–0.36	27	68	2.4	6.1
	0.36–0.40	36		3.2	
	0.40–0.44	56		5.0	
	0.44–0.48	73		6.5	
	0.48–0.52	71		6.3	
	0.52–0.56	65		5.8	
2	0.56–0.60	60	580	5.4	51.8
	0.60–0.64	61		5.5	
	0.64–0.68	55		4.9	
	0.68–0.72	52		4.6	
	0.72–0.76	46		4.1	
	0.76–0.80	41		3.7	
	0.80–1.0	156		13.9	
3	1.0 –1.2	108	329	9.7	29.4
	1.2 –1.4	65		5.8	
	1.4 –1.6	44		3.9	
	1.6 –1.8	29		2.6	
4	1.8 –2.0	20	143	1.8	12.7
	2.0 –2.5	35		3.1	
	2.5 –3.0	15		1.3	
5	>3.0	–		–	
0...5	Σ	1120	1120	100	100

[a]From Ref. 25, with H_2O = 10 mm; ozone = 2 mm; turbidity factors $T(=\beta/\lambda^\alpha)$ = 2.75, β = 0.05.

To illustrate how insolation figures can be used to evaluate possible photovoltaic system performance consider the following example. When the AM1 spectrum data of Table 2.2 are considered along with the absorption and electronic properties of solar cell materials it transpires that the maximum possible photovoltaic conversion efficiency is close to 25% (see Chapter 3). A more practical figure, applicable to present 20 cm^2 Silicon solar cells, would be around 12% and if such cells were incorporated into a 1 m^2 panel with a packing factor of

80%, then such a panel when tilted at the local latitude could provide an annual output of 245 kWh at Phoenix and 167 kWh at Cleveland, to take two significantly different insolation cases cited in Table 2.1. Taking these panels as building blocks for an appropriately sloped and oriented rooftop array on a typical single-family dwelling, e.g., roof area 107 m^2 (24 X 48 ft^2), the overall packing factor might drop to around 75% and the effective cell conversion efficiency to about 10% on account of mismatching between the panels.[27,28] Under these conditions the annual electrical output of the arrays would be 20,542 and 14,002 kWh for Phoenix and Cleveland, respectively. With a power conditioning efficiency of 85% this translates to an available annual electrical energy of 17,461 kWh in Phoenix and 11,901 kWh in Cleveland. Typical electrical loads for all-electric residences in these areas have recently been calculated.[29,30] Fixed daily profiles were assumed for diversified loads (lighting, appliances, hot water) resulting in a daily demand of 35 kWh. Hourly weather data were used to establish variable profiles for heating and cooling loads, and demands were supplied via a heat pump augmented with electrical resistance heating. For Phoenix the annual total demand was calculated as 20,692 kWh, whilst the figure for Cleveland was 24,871 kWh. Thus the photovoltaic power systems considered would lead to annual energy displacement factors of 0.84 for Phoenix and 0.48 for Cleveland. Some form of energy storage would be necessary to actualize these displacement factor figures, unless there were profound and unrealistic changes in modern lifestyles that allowed electrical demand and solar supply to be equal and in phase at all times! For Phoenix the circumstances are quite favorable as peak demand is in the summer months on account of air-conditioning requirements. For a lower domestic demand than used above Backus[31] has indicated that a storage unit capable of meeting the mean demand for four days would be sufficient to allow a Phoenix home to operate independently of the utility throughout the year.

2.2 PHOTOVOLTAIC ARRAYS

A photovoltaic array is taken here to refer to the structure of panels (modules or subarrays) that houses and supports the solar cells in a photovoltaic power system. For systems designed for use under concentrated sunlight conditions the definition includes the focusing and cooling apparatus also.

2.2.1 Arrays for use in unconcentrated sunlight

Present-day modules intended for operation in unconcentrated sunlight consist of a support board (e.g., aluminum, Plexiglas, fiberglass, glass) to which the solar cells are affixed, an encapsulating covering (e.g., glass or polymeric) and usually a supporting frame so designed to facilitate connection of modules into a closely packed array. Modules for spacecraft need, in addition, to provide some form of ultraviolet and shorter wavelength filtering, and to be able to withstand micrometeorite impact and the stresses induced by repeated deep thermal cycling. These factors are not such important considerations for terrestrial arrays, but instead the construction must contend with some or all of the following conditions:[32] humidity, air pollution, bird droppings, high winds, salt spray, cloud, fog, the impact of rocks, hail, sand, rain and snow, handling during transportation and integration, vandalism and aesthetic considerations. Birds can be discouraged from settling on arrays by the simple expedient of incorporating several short rods along the perimeter of modules, but the climatic factors turn out to be very demanding of the encapsulant, especially as the capital-intensive nature of photovoltaic power systems requires component lifetimes of at least 20 years. It is possible that no single encapsulant will provide adequate protection against all conditions and the most suitable material for a particular mission will have to be selected on the basis of the expected values and combinations of the climatic variables at the location in question. Other factors influencing encapsulant selection include optical properties and material properties related to mechanical and thermo-mechanical strength. These factors can be assessed by short-term measurements, but the durability of the encapsulant in real-weathering requires extended exposure to natural environmental conditions, and because of the infancy of terrestrial photovoltaics this information is not yet available. However, some potential candidate materials have already been identified,[33-35] e.g., for covers—soda-lime glass, low expansion coefficient borosilicate glass, acrylic, FEP Teflon, Saran, Tedlar and Mylar; for sealants—acrylic, polyvinyl butyral and poly-sulfides; for adhesives—

acrylic, epoxy, FEP Teflon and methyl phenyl types of RTV silicones. As an alternative to the glued-cover approach to encapsulation, integral covers might be considered. One scheme under investigation uses a glass fiber reinforced plastic on an acrylic base and appears to be satisfactory provided complete polymerization of the acrylic is achieved during the bonding process.[34] The borosilicate glass, Corning 7070, also seems suitable as an integral cover material and the development of an electrostatic field-assisted glass to silicon bond technology shows promise for mass-production.[36] Array assembly costs of as little as 10¢ per peak watt have been predicted for this process at a production level of $100 \text{ m}^2 \text{ h}^{-1}$, i.e., about $2 \times 10^5 \text{ m}^2 \text{ yr}^{-1}$, assuming single shift operation.

It is clearly desirable that the encapsulation process should be compatible with a high production capacity, and it would also be advantageous if there was some degree of standardization in the products of various manufacturers. Standardization as regards module physical dimensions and electrical interconnections should not be difficult to obtain and will no doubt accompany the establishment of arrays for large-scale systems, but in the present small-scale and diverse market there is considerable variation in module size. Standardization of module electrical properties is a more difficult problem, which will be exacerbated by the arrival of different semiconductor materials and fabrication processes on the production scene. Even with modules containing single-crystal silicon cells the electrical characteristics vary considerably from manufacturer to manufacturer and the situation is not helped by the different nameplate ratings that are used for modules intended to perform similar functions. Power and voltage output at the maximum power point for standard conditions of illumination and ambient temperature are minimum specification requirements. Even then similarly-rated modules may well perform differently in the field due to varying thermal characteristics: this is illustrated in Fig. 2.7.[37]

For a module of given exterior dimensions and comprising solar cells of given conversion efficiency and operating characteristics, the electrical output will depend upon the number of cells and their method of interconnection. In order to obtain the desired operating current and voltage values a combination of series-parallel connections is usually necessary either at the cell level within a module or at the

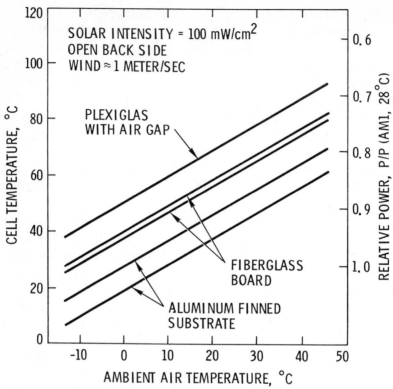

Fig. 2.7. Operating temperature of solar cells when mounted on various support boards. (After Ross;[37] courtesy of IEEE)

module level within an array. A reliable photovoltaic system would include series-parallel interconnections at both levels so that the shading or failure of one cell would not reduce the module output power to zero, as would be the case if all the cells were connected in series. Shunting of series-connected cells with protective diodes that become forward biassed if a cell develops an open circuit is possible, but would add to the cost and assembly of the array.

As individual solar cells get larger and larger, e.g., 15 cm diameter wafers or 10 X 10 cm squares (Chapter 4), the collected photocurrent per cell increases correspondingly, providing the resistance of the collecting grid does not lead to excessive I^2R losses. To minimize this loss yet maximize the cell active area (area between the grid contacts

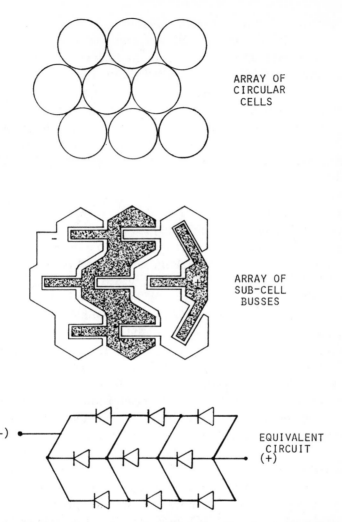

ARRAY OF
CIRCULAR
CELLS

ARRAY OF
SUB-CELL
BUSSES

EQUIVALENT
CIRCUIT

(−)

(+)

Fig. 2.8. Novel interconnect scheme for solar cells. Six wires contact the top of
each cell to the larger of the two busses below it. A suitably-cut thin
insulating film isolates the back of each cell from the larger buss below
it but allows contact to the smaller buss. (After Pryor et al.;[38] courtesy of IEEE and Motorola Semiconductor Products, Inc.)

exposed to the illumination) it is convenient to divide the cell area into a plurality of smaller cell segments, each in parallel with the other. This then allows transfer of the burden of summing the currents from the front surface metallization to external busses, even though multiple interconnections between neighboring complete cells are then required. More interconnections between cells would perhaps be expected to lead to a reduction in cell packing factor, although this can be minimized by suitable wraparound contact design. Another recent scheme utilizes copper sheets for the busses which are cut to suitable geometries, placed under the cells and separated by a thin insulator.[38] Figure 2.8 illustrates the method for both series and series-parallel connections, with six top contacts being made from each cell to the buss beneath it. The thicknesses of the buss sheets can be increased in regions of high current in order to reduce the voltage losses that would otherwise occur along the connections between parallel cells.[39] The circular cells shown in Fig. 2.8 are arranged in a hexagonal close-packed or staggered-circle fashion and this is clearly optimum for this cell geometry. With the use of half-cells (connected to the parallel strings) around 88% of the area of a square module can be rendered active, provided the module is made large enough, Fig. 2.9.[37] Further improvement on this can be expected with the advent of high efficiency hexagonal or rectangular cells.

By increasing the module size the ratio of border area to active area can also be reduced and so improvements made in the overall array packing factor. The utilization of large modules might also represent some savings as regards installation costs, although to be counted against this is the fact that replacement operations might be hazardous due to high module output voltages and difficulty in turning off the array power. Also, as more cells are incorporated into a module the effects of a single cell failure become less tolerable and careful design of series-parallel configurations becomes necessary to ensure that heating of the defective cell, due to absorption of power from the rest of the module, does not cause further deleterious effects.[37] The more parallel-connected cells the better in this regard, suggesting that future modules may be rated at high current and low voltage levels and then series connected to give higher array voltages.

Recent estimates indicate that 75% of the present cost of silicon solar cell arrays stems from the cost of the solar cells themselves.[40] Re-

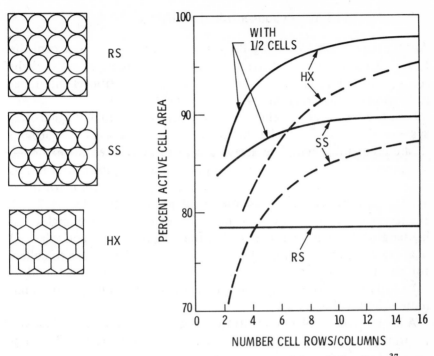

Fig. 2.9. Solar cell packing efficiency for square modules. (After Ross;[37] courtesy of IEEE)

ducing solar cell costs is thus a necessary requirement for any large-scale development of photovoltaic power systems and the progress being made in this direction is treated elsewhere in this book. At first lower cost cells will almost inevitably mean less efficient cells and thus larger arrays for a given output power. This could present a siting problem especially in urban point-of-use applications where space is particularly limited. Otherwise co-location of supply and utilization makes much sense for houses, apartment blocks, office buildings, shopping centers and light-industries as existing surfaces in the form of roofs, walls or covers for parking lots provide a ready means of multiple land-use. A displacement of conventional roofing materials also has economic attractions. At the central power station level deserts, mountainsides, sheltered waterways and existing dams are prospective locations.

2.2.2   Arrays for use in concentrated sunlight

Another approach to solving the array cost problem is to accept the high expense of the solar cell item and to incorporate cheap sunlight concentrators into the array, so increasing the power level and hopefully reducing the dollar per watt figure for the array.

Simple concentration schemes, such as the sawtooth and egg carton systems shown in Fig. 2.10 that provide concentration ratios around 2, are compatible with the arrays described above; see for example, the proposed photovoltaic power supply for a shopping center in Los Angeles where an array tilted at 45° and a north-facing reflector tilted at 35° apparently enhances the incident insolation by 19.2%.[45] Higher values of concentration imply smaller acceptance angles and higher absorber temperatures so that tracking of the sun and solar cell heat rejection become important considerations. This leads to collector designs considerably different from those just discussed, and to solar cells suited to operation at high temperatures and in high illumination fields (Chapter 5).

A list of typical solar concentrators is given in Table 2.3 and all are, in principle, applicable to photovoltaic power conversion. The con-

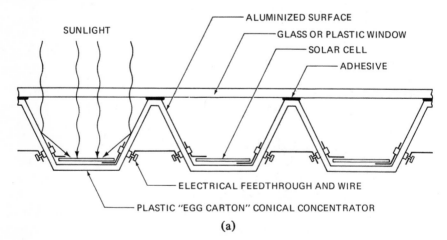

(a)

Fig. 2.10. Some solar concentration schemes suited to photovoltaics. (a) egg-carton (After Cherry;[41] courtesy of Amer. Soc. Mech. Eng.); (b) sawtooth; (c) dielectric compound parabolic (Adapted from Gorski et al.[42]); (d) floating eyeball (After Mash[44] with permission); (e) fixed mirror trough (Adapted from Backus[43] with permission).

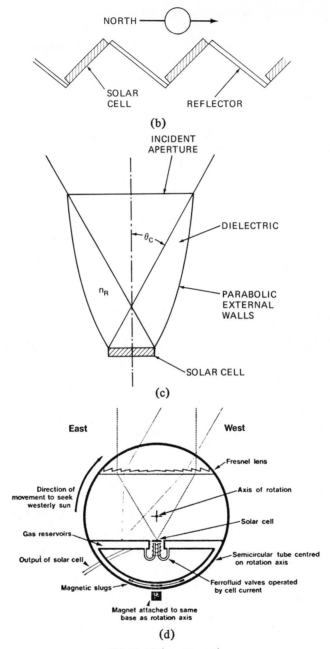

Fig. 2.10 (*continued*)

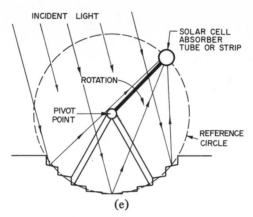

(e)

Fig. 2.10 (*continued*)

centration ratios quoted in the table refer to the average energy density on a fixed surface normal to the sun. An alternative definition is[42]

$$C = \eta_{op} \frac{a_o}{a} \tag{2.1}$$

where a_o is the incident aperture area, a is the solar cell area, η_{op} is the total optical efficiency of the concentrator-cell combination and includes all reflection and transmission losses. Values of C computed from Equation 2.1 may differ somewhat from those arrived at via energy density considerations, but the ranges of figures quoted in Table 2.3 are still appropriate.[46] Equation 2.1 when applied to ideal non-imaging concentrators leads to[46]

$$C_2 = \frac{n_R}{\sin\theta_c} \quad \text{and} \quad C_3 = \frac{n_R^2}{\sin^2\theta_c} \tag{2.2}$$

for the cases of two-dimensional and three-dimensional optical systems respectively, where n_R is the refractive index of the concentrating medium and θ_c is the acceptance angle. Practical compound parabolic concentrators that come close to realizing ideal concentration values have been developed and show considerable promise for photovoltaic applications.[42,46,47] Figure 2.10c illustrates an all dielectric compound parabolic concentrator in the form of a trough with parabolic sides whose axes are inclined at $\pm \theta_{c\,max}$ to the optic axis and whose foci are at the base of the opposite sides. The end walls can be

Table 2.3. Typical Solar Concentrators.[a]

NAME	METHOD OF CONCENTRATION REFLECTIVE	REFRACTIVE	USUAL RANGE OF CONCENTRATION RATIO	TYPE OF TRACKING REQUIRED NONE	ONE AXIS	TWO AXIS	FOCAL ZONE POINT	LINE	AREA
Flat reflectors									
Side mirrors (north side of the absorber, noon reversible mirrors, "V" troughs, etc.)	X		1.5–3.0	X					X
Fixed flat mirrors, movable focus	X		20–50		X (absorber)			X	
Multiple heliostats redirecting to a central absorber	X		100–1000			X	X		
Single curvature reflectors									
Truncated cones	X		1.5–5	X	X		X		
Compound parabolic concentrator	X		3–10	X	X		X	X	
Parabolic cylinder (E–W, N–S, or tilted axis)	X		10–30		X			X	
Reflecting linear Fresnel	X		10–30		X			X	
Double curvature reflectors									
Paraboloids	X		50–1000			X	X		
Hemispheres	X		25–500			X	X		
Reflecting circular Fresnel	X		50–1000			X	X		
Refracting lenses									
Linear Fresnel		X	3–50		X			X	
Circular Fresnel		X	50–1000			X	X		

[a]From Ref. 21.

inclined to provide additional concentration. The use of total internal reflection at the dielectric interface rather than traditional mirror-reflection in an air-filled system reduces fabrication problems and deterioration in reflecting properties, besides allowing either an increased concentration ratio for a given θ_c, or the same C for a larger θ_c. Radiation funnels of this design have been fashioned in acrylic ($n_R = 1.5$) by injection molding techniques and an arrangement of 108 such units is shown in Fig. 2.11a. Annealing of the acrylic bars before bonding to specially designed Si solar cells (0.284 cm $\times$ 2.51 cm, $\eta = 12\%$) apparently improves the subsequent bond, which at present, is made using a room-temperature vulcanizing clear silicone rubber. Aluminum extrusions form an extended surface heatsink, this feature being a particular attribute of refractive-optics systems which cannot be matched by reflective-optics systems in which the heat rejection surface must be kept to a minimum so as not to shade the reflector excessively.[27] Modules giving a concentration ratio of 6.11 and η_{op} value of 82% have already been constructed, and this scheme is attractive not only on account of its production simplicity, but also because its uniform response over a wide acceptance angle ($\simeq 10°$) makes diurnal tracking unnecessary.

Two systems that have utilized refractive optics to attain high concentration ratios are shown in Figs. 2.11b and 2.11c. The unit shown in Fig. 2.11b is a 100 W demonstration unit array built in the United States by R. C. A. and employs vinyl plexiglas lenses to give a concentration ratio of 330. The system shown in Fig. 2.11c is rated at 1kW$_e$ and uses 135, 30.2 cm $\times$ 30.2 cm, circularly-patterned Fresnel lenses to concentrate sunlight onto 135 Si cells of area 15.2 cm^2, giving a geometric concentration ratio of 60.[27] Square lenses allow a simple support structure, although hexagonally shaped lenses would give better uniformity. Again an extended Al heatsink is utilized and a Dow-Corning RTV bonding material (3140) attaches the cells and provides electrical isolation. To perform the latter function and also provide good thermal conductance is a demanding criterion for the adhesive and may necessitate amelioration by the use of anodized aluminum,[27] or a thin layer of insulating plastic, e.g., Mylar.[42] Other problem areas that have come to light as a result of operating experience with this system, and which are likely to be present in other concentrator schemes, concern cell interconnections and optical alignment.

Low resistance connections are particularly necessary in concentrator arrays because of the large generated currents, but rigid electrical bonds that might help achieve this could lead to considerable thermal expansion mismatch and ultimate failure. Flexure joints and utilization of aluminum or copper interconnects seem to be required[48] along with, perhaps, braided buss straps to allow some freedom for individual cell alignment with the associated concentrator. The projected maximum power for the above array is 729 W for the passively cooled case with no fins on the heatsink (cell temperature = 93°C), and 857 W for a water cooled case (flow rate = 0.45 kg sec^{-1}, cell temperature = 60°C).[27] Heat rejection at these power levels thus seems to demand active cooling, and some basis for comparing various cooling schemes and bond resistances is provided by the calculated data for Fig. 2.12.[49] Passive cooling to ambient air when using extended surfaces would seem to be satisfactory for the radiation levels achievable with practical linear, trough or compound-type concentrators. The efficacy of back surface heat removal indicates that the use of optical filters to shield cells from infrared radiation, or cell immersion in coolants such as flowing water, may not be necessary except at the highest concentration ratios. Besides bringing about cell cooling, immersion in a liquid may cause optical-absorption, electrical shunting and effects due to the presence of a semiconductor-electrolyte interface.[50]

For high concentration ratio systems sun tracking is required and in the experimental system described above (Fig. 2.11c), two-axis tracking is provided by an azimuth/elevation mount. This is controlled by a photocell sensor and a cloud detector which prevents the array from seeking reflecting surfaces other than the sun when operating in cloudy conditions. Active tracking is preferable to preset timer tracking, owing to the variation in apparent solar trajectory throughout the year. Two-axis tracking is also used on another experimental concentrator system rated at 1 kW$_e$ that uses 120 GaAlAs/GaAs solar cells and conical concentrators with mirror surfaces.[51] This expensive concentrator method accounts for 80% of the system cost which has led to the consideration of cheaper materials such as slump glass and aluminum for the concentrators.[3] Another presently expensive, but ingenious, concentrator scheme involves a 25 cm diameter Fresnel lens inside a 30 cm diameter hermetically sealed plastic sphere which also

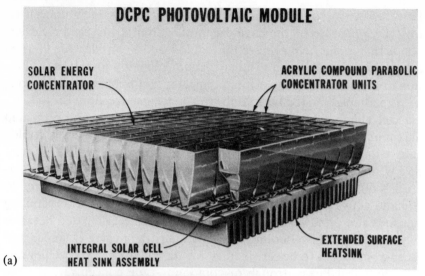

Fig. 2.11

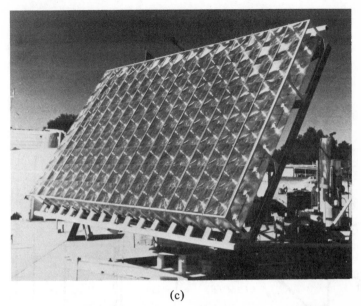

(c)

Fig. 2.11. Experimental photovoltaic concentrator arrays. (a) 100 W dielectric (acrylic) compound parabolic concentrator;[42] (b) 100 W vinyl plexiglass lens concentrator; (c) 1 kW Fresnel lens concentrator.[27] Photographs courtesy of A. Gorski, Argonne National Laboratory; M. Nowogrodski, RCA Labs; E. L. Burgess, Sandia Labs (with acknowledgement to ERDA).

contains a GaAlAs/GaAs cell and four gas-filled reservoirs, Fig. 2.10d.[44] Off-axis sunlight causes gas in one of the reservoirs to expand, so moving a magnetic slug and rotating the whole assembly against the field of a fixed external permanent magnet. High manufacturing costs and difficulties in controlling movement during periods of intermittent sunshine would seem to render this scheme unsuitable for large-scale use. The same cannot be said for another concentrator scheme under development which is unusual because of the fact that it utilizes a movable absorber and a fixed concentrator,[21,43,48,50] thus greatly simplifying the steering structure. Such a fixed mirror concentrator is shown in Fig. 2.10e, where the reflector is comprised of a series of long, narrow strips of flat mirrors along a cylindrical trough. The faceted nature of the collector introduces some optical losses but apparently these can be reduced to below 10%,[48] and so do not negate the unique feature of this design,

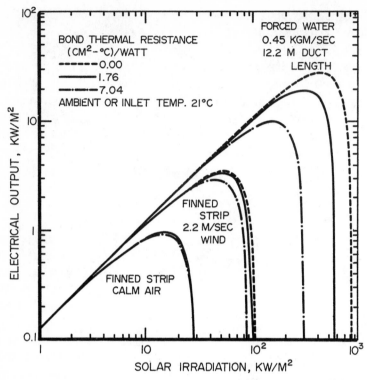

Fig. 2.12. Calculated effect of solar cell/heatsink bond thermal resistance on the output of differently cooled and mounted silicon solar cells as a function of solar irradiation. (After Florschuetz;[49] courtesy of IEEE)

namely the provision of a sharp line focus on the reference circle of the collector cylinder for any incoming light direction. The fixed nature of the mirrors means that large sizes can be contemplated (perhaps 10–15 m from rim to rim[43]), enabling absorber diameters of around 20 cm and concentration ratios of about 20–50. Solar cells in a tubular form, which would facilitate forced-cooling, have been fabricated and could be useful in this concentration scheme.[52] However, conventional flat cells can also be used, and even be expected to perform satisfactorily at these concentration levels using passive cooling with finned or pinned surfaces and assuming reasonable wind conditions (Fig. 2.12). The strip geometry of the absorber for this system suggests the use of a series string of solar cells but this is not practicable because of shading effects. A parallel-series connection of solar cells is required to prevent unacceptable fluctuations in module output,[50] and will perhaps necessitate the use of narrow cells similar to those incorporated in compound parabolic troughs.[47]

Shading of concentrators by neighboring units is another factor to be considered in choosing a suitable concentrator scheme. The optimum arrangement for a field of modules must take into account land costs and topography as well as module placement; a universal optimum cannot be specified because of the dependence of module orientation for maximum insolation on local latitude. There are indications that a minimum cost-of-energy figure can be arrived at for various concentrator systems,[50] but some shading of any given module has to be expected so that an entirely series-connected array of modules should be avoided.

Photovoltaic power generation is still in its infancy and schemes utilizing solar concentration are particularly so. The experience that will be gained through constructing and operating the above schemes and others will be very valuable and should provide much of the data required to assess the relative merits of concentrator and flat plate systems more critically. It is not yet possible to make meaningful cost estimates of practical systems but Sandia Laboratories in the United States are hoping for a capital cost of only $3.5/peak W for their system,[27] and projected costs for fixed mirror solar concentrators are about $10-20/m^2,[43] which is about two orders of magnitude cheaper than that of the solar cells they might currently replace. Sunlight concentration thus appears very promising and photovoltaic systems using conventional cells and low concentration ratio collectors may well figure prominently in intermediate term (1980–1990) markets. The static collector systems, particularly the compound parabolic concentrator, could be used in rooftop point-of-use systems and co-located intermediate power level applications. The steerable arrays with concentration ratios of 100–1000 are more suited to central power station operation.

2.3 POWER CONDITIONING AND SOLAR/NONSOLAR POWER PLANT INTERCONNECTION

For applications where photovoltaic power systems are required to supply a predictable and small (less than a few kilowatts) load, a simple direct battery charge system is usually adequate, e.g., Fig. 2.1. A series blocking diode prevents discharge of the battery through the solar cells during low insolation conditions, and a shunt regulator prevents overcharging of the batteries that might otherwise occur during periods of prolonged insolation. The strategy here is to choose an array and battery size capable of supplying the load on demand and to waste any power that might on occasion be generated in excess of the demand (load plus battery charging). The application of this concept to larger loads at the residential (5–10 kW) and intermediate (0.1–10 MW) levels might not be acceptable for at least two reasons. Firstly, the sheer area of collector required for a self-sufficient supply is likely to be prohibitive, especially in the case of commercial and industrial loads. For residential supply, even in such a favorable location as Phoenix, Arizona, 100% conventional energy displacement is hardly feasible with acceptably sized roofs (Section 2.1). Secondly, the electrical sizing is such that it would be difficult to rationalize the amount of energy that may have to be dissipated wastefully. A more practical photovoltaic power supply at this level would not therefore be expected to stand alone and would incorporate a backup supply from a nonsolar generator source. For a residential supply in a high-insolation area this backup might be in the form of a diesel generator so that independence from the conventional electrical grid could be achieved if so desired. However, interconnection with the local utility might be more acceptable to many consumers, and would certainly be the case at intermediate load-level applications. At the central power station level (50–5000 MW) interconnection with the conventional grid can be taken for granted. The power flow strategy for these cases where co-located energy storage is available would be to use the photovoltaic power to supply the load first and then to use any surplus for charging the storage facility, with any extra load being met by the utility.

Three possible interconnection schemes are shown in Fig. 2.13. The direct charge system would just be a scaled-up version of the system shown in Fig. 2.1, with the provision of an inverter for the supply of ac loads and to match the input from the utility which would otherwise need to be rectified. Preservation of the dc nature of the photovoltaic power for the supply of some portions of the load might be considered, e.g., in houses, offices or shopping centers with substantial air-conditioning requirements that could be met by dc motors driving fans and pumps.[45] However a mixed ac/dc supply is likely to complicate the power control situation. Both solid-state and electro-mechanical inverters are available with ratings up to megawatt levels and dc to ac conversion efficiencies approaching 95%. The compact and essentially maintenance-free solid-state units would be particularly suited to residential usage. The circuits in Figs. 2.13b

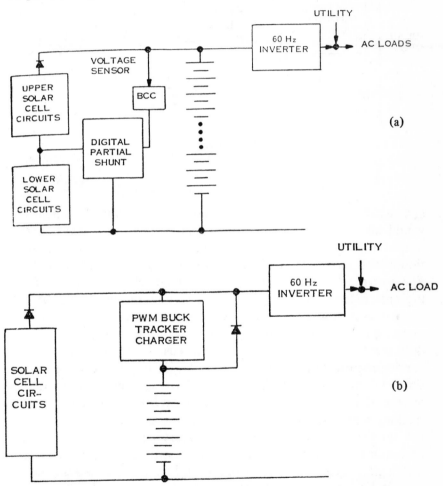

Fig. 2.13. Photovoltaic power system interconnect schemes. BCC is the battery charge control sensing unit and PWM indicates maximum power point tracking via pulse width modulation techniques. (After Shepard et al.;[30] courtesy of IEEE)

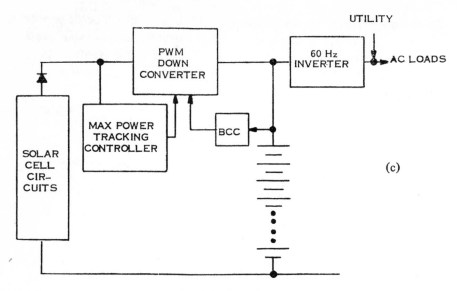

Fig 2.13 (*continued*)

and 2.13c contain a maximum-power-point tracking element which would take the form of an electronic arrangement to sense the array output and provide the input impedance required to maintain operation at the maximum-power-point irrespective of the insolation level. The provision of the binary-segmented digital shunt in Fig. 2.13a prevents battery overcharging and this function could also be performed, as is shown in the other circuits, by using a pulse width modulated converter which drives the array towards open circuit conditions when the available power exceeds the total demand.[30] No maximum-power-point tracking is shown in the direct charge circuit since, in this case, essentially constant voltage operation is dictated by the floating battery. Operational experience with the various tracking schemes is needed to assess their usefulness,[299] and continuous tracking may not always be necessary because for an array the optimum load resistance is not nearly as critically defined as it is in the single cell case. This is illustrated in Fig. 2.14 and arises from the mismatch, as regards electrical properties, between cells constituting an array. However, this effect does cause an overall degradation in conversion efficiency and so, in practice, arrays should comprise near-identical cells.

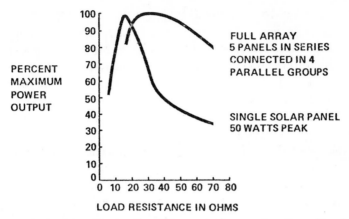

PERCENT
MAXIMUM
POWER
OUTPUT

FULL ARRAY
5 PANELS IN SERIES
CONNECTED IN 4
PARALLEL GROUPS

SINGLE SOLAR PANEL
50 WATTS PEAK

LOAD RESISTANCE IN OHMS

Fig. 2.14. Normalized cell and array power as a function of load resistance. (After Haas et al.;[53] courtesy of IEEE).

If the rationale for interconnection of photovoltaic power plant and utility is accepted then a logical extension of the above systems would be to allow for the feedback of otherwise wastefully-dissipated photovoltaic power to the utility. This power, provided it met the required standards as regards voltage and harmonic content, would doubtless be acceptable to the utility as it could be used elsewhere in the grid and would lead to some reduction in conventional fuel consumption. Such a scheme might pose some institutional problems, prominent amongst which would be the questions of payment for the fedback power and ownership of the photovoltaic power plant. Proprietor ownership would probably lead to the same power handling strategy as discussed above, but with utility ownership the possibility of rationing the stored power for release only at peak demand time arises. In this manner the photovoltaic storage system would make an impact not only on conventional fuel consumption but also on conventional peaking and intermediate-level power plant generating capacity. Because of its zero fuel and low operation and maintenance costs the photovoltaic power plant would be kept in service at its maximum possible output for as long as possible during the day and would assist the nonsolar plant in supplying the demand and charging the energy store. Charging could be completed at night when the consumer demand is minimum, thus leveling the nonsolar generator load and allowing peak-shaving. Discharge of the store would be allowed to continue as long as this was cheaper than adding other units (nonsolar) into the system and there was sufficient energy in the store. This scheme is clearly more attractive economically than one in which the storage facility is dedicated to providing power during cloudy periods.[54] If, then, the stored energy is to be used at the discretion of the utility it seems practical to consider locating the storage units on utility premises, either at nonsolar power plant sites or community storage substations.

The merits of this approach are several, not least of which is the removal of operating and maintenance responsibility from the possibly uninterested and unqualified consumer. Even storage batteries need some attention; their central location would facilitate this and remove the need for the establishment of correctly ventilated and insulated enclosures on the consumers' premises. A common storage pool could utilize larger rated, and thus usually cheaper, units than would be necessary for individual consumers and also allow consideration of alternative storage schemes to the relatively simple battery approach. Any conditioning of the utility power that is required for feeding the energy store, e.g., ac/dc conversion for battery or electrolysis storage, can be more conveniently performed at a central large facility, and would also simplify consumers' power conditioning plant. The same argument applies to the location of power control equipment that would be required to maintain optimal power flow.

The somewhat speculative nature of this section attests to the dearth of information and experience relevant to the composition and operation of other than small, independent photovoltaic power systems. Only as larger systems are commissioned can the decisions be made as to the relative merits of: maximum-power-point tracking versus battery floating; independent photovoltaic systems versus systems interconnected with nonsolar power plants; and systems possessing no energy storage versus systems with storage that is either co- or centrally-located.

2.4 ENERGY STORAGE

Almost all the small, point-of-use photovoltaic power systems that are in use today incorporate an energy storage element to provide power during periods of inclement weather and at night. Storage is via secondary batteries which provide a simple system having minimal environmental impact and needing little maintenance. At higher power levels appropriate to residential, commercial and central station systems and where there are tie-ins to nonsolar power plants, the question of storage is not so simply dealt with, as was indicated in the preceding section. Storage is generally considered necessary for useful operation of solar power systems, but this may not be the case when substantial loads occur during daylight hours of reliable sunshine. Under these circumstances the solar plant could be considered as operating in an intermediate plus peaking mode, thus serving to reduce the need for added nonsolar generating capacity and energy storage. Even when the insolation pattern does not allow this mode of operation the capital-intensive, zero-fuel cost nature of photovoltaic conversion would decree that the solar electricity be dispatched whenever available and so energy storage considerations from a utility standpoint might not be changed much from the current attitude.

The present evaluation of energy storage can be expressed in terms of credits which apply to cost savings in utility capital and operating requirements resulting from the capability of the storage device to reduce peak loads on generation, transmission and distribution equipment. For example, credit would be given for obviating the need for peaking and intermediate load units to be started up for short intervals and then shut down; for replacing spinning reserve capacity; for being able to follow rapid load fluctuations and thus allowing conventional thermal units to operate at steady or slowly varying outputs; for being able to operate in dispersed locations so allowing deferment of additional network capacity, consumer-protection against transmission and distribution outages and maintenance of existing substation fault current levels.[55] One additional credit that could result from having a non-conventional input to the electrical system would be the contribution to security of supply at the generation level. Thus from the utility's standpoint the presence of a photovoltaic plant is likely to increase the attractiveness of energy storage, and this would be true even with point-of-use photovoltaic systems provided the dispersed storage scheme outlined in the preceding section was adopted. The questions that now need to be answered are what storage methods are available and which ones are suited to particular locations, i.e., at the central station, at a dispersed station or at the point-of-use? In answering these questions the appropriateness of the storage method to storing photovoltaic electricity will be considered even though, as mentioned above, in large systems containing a mix of photovoltaic and nonsolar plants it may be more expedient to charge the store from the nonsolar generation component.

A summary of possible electricity to electricity storage schemes is given in Table 2.4. The overall conversion efficiency figure that is quoted cannot be used to rank the storage schemes unequivocally as any meaningful economic analysis must extend over the entire electrical system and also consider the likely effective storage plant load factor.[56] It is possible, for example, that storage devices operating at relatively high conversion efficiency and low load factor will have the same break-even capital costs as storage plants operating at low efficiencies and at high load factors.[56] For low load factors (< 0.1) the increment in acceptable capital cost per unit improvement in conversion efficiency apparently becomes very small after about the 55% figure, but does not show similar saturation at load factors in excess of about 0.2. Thus there is an economic incentive to seek high conversion efficiencies if storage plants can be viewed as providing both peaking and intermediate-level power; photovoltaic-supplied storage plants could fall into this category. On the other hand, the long-sequence and thus low efficiency hydrogen scheme may also prove to be very attractive because of the versatility of the intermediate product. Besides its use in fuel cells, hydrogen can be used for heating, transportation (via internal combustion engines), synthetic fuel production and is also conveniently transportable via pipeline or tanker.[57]

The selection of a particular storage scheme is thus seen to be a nontrivial matter and the situation is particularly unclear at the present time because of the experimental and/or untried nature of many of the methods. Specific comments on the various schemes are given below.

1. Batteries are the obvious choice for storage of photovoltaic electricity because of the simplicity of conversion and compatibility of terminal voltages. Present-day photovoltaic systems usually employ lead-acid batteries at a cost of around $70 kWh^{-1} and this expense can be tolerated because of the modular size which allows installation at the load center. However, the main reason for their present use is that there is no practical alternative. Such an alternative may have to

Table 2.4. Summary of electricity storage schemes.

METHOD	CONVERSION	STORAGE MEDIUM	RECONVERSION	ELEC. OUTPUT	CONV. EFFICIENCY	LOCATION: POINT OF USE	DISPERSED	CENTRAL	SEE NOTE IN TEXT
Battery	electrochemical	battery	—	dc	70–80	X	X	X	1
Pumped-hydro	ac[a] → pump	water	electromechanical (turbine)	ac	70–75			X	2
Compressed air	ac[a] → compressor	compressed air	expand + heat → electromechanical (gas turbine)	ac	65–75			X	3
Hydrogen	photoelectrolysis → gas	H₂ gas H₂ liquid H₂ slush metal hydride	gas → fuel cell	dc	35–50	X	X	X	4
	electrolysis → gas		gas → electro-mechanical (aphodid burner + steam turbine)	ac			X	X	
Flywheel	ac[a] → motor	flywheel inertia	electromechanical (motor-generator)	ac	70–80		X		5
Superconducting magnet	electromagnetic	superconducting magnet	—	dc	90–95			X	6

[a]Only necessary if dc machinery cannot be used.

be developed before any significant adoption of batteries can be expected from users of electricity at the kilowatt to megawatt levels. This is because the lead-acid battery does not meet the projected specifications of performance, cycle life and cost that utilities would demand before significantly increasing the installed capacity of battery storage beyond the level currently employed for load leveling. The techno-economic criteria can be expressed as an energy density of around 100 Wh kg^{-1}, a long life of around 2000–2500 cycles and cost of around \$20–35 kWh^{-1}. Table 2.5 enables some comparison of past and future batteries to be made.

The sodium-sulfur and lithium-sulfur systems show some promise but the required operation at 300–400°C introduces material problems and undesired energy consumption. The development of suitable materials is likely to delay the introduction of these batteries until 1980–85,[58] at which time it might be considered feasible to obtain the required operating heat from thermal energy rejected by concentrator solar cell arrays. Metal-air batteries are attractive from a materials cost point of view but little progress has been recorded in developing secondary air electrodes with long lifetimes.[58] The availability of nickel is expected to limit the widespread use of nickel-related

Table 2.5. Storage battery data.[a]

BATTERY TYPE	ENERGY DENSITY[a] Wh.kg^{-1}	LIFE CYCLES[b]	COST \$/kWh
Silver–Zinc	100–120	100–300	900
Nickel–Cadmium	33–40	300–2000	600
Nickel–Iron	22–33	3000	400
Lead–Acid[c]	11–22	1500–2000	50
Nickel–Zinc[d]	66–88	250–350	20–25
Zinc–Chlorine[d]	66	500	10–20
Sodium-Sulfur[d]	170–220	1000	15–20
Lithium–Sulfur[d]	130–170	1000	15–20
Zinc–Oxygen[d]	160	?	?
Aluminum–Air[d]	240	?	?

(a) Values shown are for 1 hr and 6 hr rate of discharge respectively.
(b) Range shown is from severe to modest duty.
(c) Typical automotive battery.
(d) Projected figures—batteries still at experimental stage.

[a] Data taken from Refs. 57 and 58.

battery systems, and the nickel-zinc cell is further hampered by the short cycle-life of the zinc electrode.[58] The zinc-chlorine system has become of interest mainly due to the recently-proposed method of storing the toxic chlorine in the form of chlorine hydrate. Operation is at room temperature and on demonstration of suitable life properties this system could be very suitable for large scale energy storage. Presently, the conversion efficiency of 80%, the cycle life of 10 years and the proven technology make lead-acid batteries the only viable units. Costs are unlikely to fall below $20–30 kWh^{-1} because of the use of lead, and in large-scale uses consideration must also be given to the cost of the copper that would have to be used in interconnecting many small units. Development of large individual batteries is thus required; lead-acid systems possess this capability with 64,000 Ah units occupying about 2.7 m^3 having already been reported.[59]

2. Pumped hydro is the only storage method presently employed by utility companies for large-scale use: in the United States, for example, the installed capacity is around 10,000 MW. Continued development of this storage method is severely restricted by the lack of suitable geographical sites at which to build storage reservoirs, particularly close to urban load centers. Some expansion may be possible through locating the power plant and lower reservoir in underground caverns, and this may also provide an extra design variable in that the differential head is not fixed by topographical features. Both modified Francis waterwheel reversible pump/turbine units, and tandem arrangements of impulse turbines and multi-stage high pressure pumps on a common shaft are being developed for use in this application.[60] The latter run in the same direction whether pumping or generating and so starting and changeover times are fast (1–2 min.), whereas with reversible units these times are about three times longer. The size and relative sluggishness of this storage system ensures its use only at central station locations and for meeting predetermined peak load demand. Conventional pumped hydro can be expected to expand as much as is environmentally possible on account of its proven practicability, and underground systems are likely to materialize, with economics being significantly dependent on credits or debits arising from excavation procedures.[61] The obvious arrangement for incorporating a photovoltaic power plant into this scheme would be to float the solar cell arrays on pontoons on the upper reservoir, thus providing a very energy-intensive locale.

3. Storage arrangements involving compressed-air—combustion turbine systems have a greater flexibility with respect to size than pumped hydro systems and also minimize the requirements for above-ground real estate. Potentially suitable storage reservoirs are natural caverns, solution dissolved salt beds, porous strata bounded by hard rock layers, abandoned mines, aquifers, nuclear excavated caverns or grids of large tunnels. The latter might be necessary to minimize gas leakage problems due to jointing, bedding and other rock discontinuities. The compressor machinery would be located above ground and if the heat liberated during gas compression could be stored this could be later used during the gas expansion cycle and thus lead to an independence from oil, which would otherwise be used for gas heating prior to electricity generation. Advanced hard rock mining techniques would have to be employed to implement this scheme and the uncertainties involved in this and other aspects of the storage method make technical and economical assessments difficult at this stage. The use of photovoltaic power plants, as opposed to conventional nonsolar plants, to drive the storage compressors does not seem to offer any advantages or disadvantages (except cost at this time, of course).

4. The electrolysis-hydrogen sequence for energy storage is highly compatible with photovoltaics on account of the low voltage dc requirements of electrolysis. Commercial electrolysers are available with conversion efficiencies (ratio of heat content of the evolved hydrogen gas to the electrical energy input) in excess of 70%. Operation at high pressures (2×10^7 Nm^{-2}) and moderate temperatures (200°C) has enabled efficiencies of 90% to be reached in experimental units.[62] Photoelectrolysis, in which light is absorbed in suitable semiconducting electrodes immersed in an electrolyte, is a potential high-efficiency process for converting light and electricity to hydrogen but, as yet, has not been developed. Hydrogen can be stored in gaseous, liquid or semi-solid form, and also as a metal hydride on reaction with appropriate metals.[57] All methods are expensive, incurring high costs due to containment, refrigeration and availability of suitable metals respectively. The latter method warrants further study as it provides a compact, nonrefrigerated store with simple hydrogen release at only slightly above room temperature. With the advent of a suitable storage medium hydrogen will undoubtedly play

a role in meeting future transportation and heating needs, but its use in electrical storage systems depends largely on the development of a cost-effective fuel cell.

Present fuel cell problems include the need to use platinum as an electro-catalyst, the ageing of electrodes (particularly due to recrystallization), the lack of data on porous electrode design,[57] and the difficulties in making large multi-cell units. Considerable development is required before fuel cell systems can begin to compete with batteries for large-scale electricity storage and reconversion. Useful information should result from the construction in the United States of a 25 MW fuel cell power plant for utility service, even though the proposed system requires noble metal catalysts and thus cannot be envisioned for extensive utilization.[63] The cell uses a phosphoric acid electrolyte but there are indications that future cells might perform more efficiently using an electrolyte of alkali metal carbonates in a ceramic matrix.[64] Such cells would also be expected to offer economic advantages as they do not require noble metal catalysts. The modular nature of fuel cells ensures some credit and, as with batteries, allows use to be considered at point-of-use, dispersed and central locations. The complex system of electrolyser–hydrogen-storage–fuel cell is perhaps too forbidding for acceptance by most residential consumers, but its situation at dispersed locations in a utility power system seems reasonable, although a limitation is the finite time required for start-up (60 min from cold for the above cell).[63] The compact nature of the components (occupying about 1% of the real estate needed for a comparable pumped hydro system)[63] would allow location close to urban areas with large demand, however for use in conjunction with photovoltaic electricity there may not be room in the latter situation to co-locate the solar cell arrays. It might then prove advantageous to locate both arrays and electrolyser away from the fuel cell and supply the hydrogen to the latter by pipeline or tanker. At central station locations an alternative to the fuel cell for converting hydrogen to electricity would be an aphodid burner,[57, 62] which consists of a hydrogen-oxygen burner with a water spray moderator to adjust steam pressure, flow and temperature prior to turning a steam turbine.

5. The coupling of a motor-generator set to a flywheel allows for the storage of electricity in the form of mechanical or kinetic energy.

The method is no more suited to operation with photovoltaic power plant than it is to use with other means of producing electricity, although the two might be linked together as photovoltaic power systems should be in widespread use by the time utility-size fly-wheels are commercially viable. The major breakthrough in flywheel design in recent years has been the development of new shapes, materials and fabrication methods to allow construction of lightweight flywheels that can be rotated at high speeds without danger of breaking apart.[65] Although utilities would likely credit flywheel storage systems on account of their ability to follow load variations rapidly, the limitations to the method are low energy density and continued energy loss due to bearing friction and windage. The latter can be minimized by operating the flywheel in a vacuum and utilizing hydrostatic or even magnetic bearings, but the energy density problem appears less tractable. For example a recent design study for a 500 MW flywheel made from high strength organic fibers in hubs of various configurations projected an energy density of about 26 Wh kg^{-1},[65] whereas a figure of 88 Wh kg^{-1} has been cited as being necessary for economic feasibility.[66] At best it would appear that flywheel systems are suited to situations requiring either rapid following of small load variations or a short lived supply during the switching between generating units.

6. Electric currents can persist indefinitely in appropriately-cooled superconducting materials, thus suggesting yet another method of electrical energy storage. The method is particularly compatible with photovoltaics because the superconducting magnet is a high-current, low-voltage dc device. The overall conversion efficiency is the highest of all the methods reviewed, but the method will probably be the last to be implemented for large-scale electricity storage. This is because superconducting magnetic energy storage (SMES) presents a series of entirely novel engineering design and construction features and, mainly on account of costs associated with refrigeration, only appears to be economical at the 100–10,000 MWh level.[67] A magnet capable of storing this amount of energy would have a radius of 50–100 m, and to resist the large forces exerted on the conductor windings it would perhaps have to be mounted in subterranean tunnels in bedrock. One magnet design suggests a conductor consisting of 6240 TiNb filaments embedded in aluminum and wound into a

single layer solenoid.[67]   Cooling would be from a pool of superfluid helium at 1.8°K. In addition to the areas of housing, conductor design and refrigeration, further study is required concerning matters of electrical insulation at 4°K, the properties of structural materials in the temperature range of 4-300°K, electrical connections at the 100,000 A level and the biological effects of huge magnetic fields.[68] Only when these technical and environmental problems have been overcome can SMES be contemplated and the potential credits of high conversion efficiency and millisecond response time perhaps realized.

2.5 PRESENT-DAY PHOTOVOLTAIC POWER SYSTEMS

Applications for which photovoltaics is presently considered a viable power source are listed in Table 1.2, and mostly can be characterized as being located in remote areas and consuming less than 1 kW of low voltage (< 50 V) dc power, generally to charge storage batteries. An example of a typical system is shown in Fig. 2.1 and besides the load, solar cells and storage battery, comprises a blocking diode and a battery overcharge protection circuit.   These last two features will usually be necessary although in some circumstances it may be possible to allow battery overcharge with impunity e.g., in vented batteries, overcharging only causes a loss of water from the electrolyte and this would be replaced from a suitable fluid reserve.[19]

The design of a photovoltaic–electricity storage power system that is to act independently of any nonsolar generating plant involves calculations of array size and battery capacity, with a superimposed uncertainty in the form of the insolation.   Because the location is likely to be remote, precise insolation data is not likely to be available and interpolation through nearest weather stations may not be valid, especially in coastal and mountainous regions.   By using data from weather stations in similar climatological zones some estimate of the insolation can be made, which then needs to be modified via a "variability factor" to allow for the variation from year to year both in mean and worst-case values.   By taking a large variability factor, say 15%,[19] it is reasonable to work with monthly insolation averages as storage capacity will be high and so system performance is unlikely to be seriously affected by short-term (daily) fluctuations.   For maximum solar enegy collection a good rule-of-thumb guide for fixed array orientation is to tilt at an angle equal to that of the local latitude, although for areas with large differences between summer and winter insolation it may be more expedient to fix the tilt angle at a value appropriate to maximum collection in winter. In this manner the input power can be levelled somewhat. An extra uncertainty introduced by tilting stems from the collection of ground-reflected sunlight and so allowances should be made for any significant changes that might occur; for example, during snowy conditions.

In addition to insolation factors it is necessary to take into account likely array and ambient temperatures and their possible variation throughout the year. Useful current can be delivered to the load and to charge the battery only if the array voltage exceeds the battery voltage plus the blocking diode voltage drop. The worst-case condition from this point of view will occur on a hot day (low array output voltage) when the battery is fully charged. This could be an autumn condition for many areas in the United States,[19] however it does not represent the worst-case as regards supply to the load, as this can be achieved via the fully charged store. The strongest factor influencing the selection of battery storage capacity is thus the occurrence of periods of continuous low insolation, and these are likely to arise in winter. The array size is primarily determined by the mean insolation level and the required load but, of course, it is not possible to compute array size and battery capacity entirely independently. Some computer calculations and design strategies have been reported[19, 69] and, to gain some insight into the system interdependencies, consider the following example in which the size of a simple photovoltaic power system to supply a daily load of 100 W at 24 V in Vancouver, Canada is calculated. The relevant solar radiation statistics for a south-facing flat plate inclined at various angles are given in Table 2.6.

Consider that the array comprises two subarrays of parallel connected modules with the two subarrays connected in series, each module having characteristics as shown in Fig. 2.15. At maximum sunlight intensity each module operates at 16 V to supply a battery charge voltage of 30 V after allowance for a 2 V drop across the blocking diodes. For a cell temperature of 25°C the module output current at this voltage is 700 mA for 100 mW cm^{-2} radiation. The

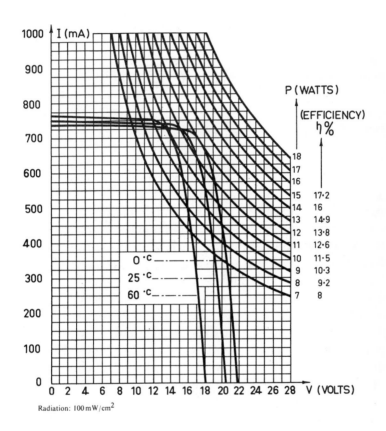

Fig. 2.15. Characteristics of Philips BPX47A silicon solar cell array.

minimum array size A that can meet the load requirements can be found from

$$A = \frac{2l}{(0.9 \times 0.97 \times 0.85)} \times \frac{100}{0.7 \times D_m} \qquad (2.3)$$

where l is the load in Ah, D_m is the mean annual insolation in mW h cm^{-2} day^{-1}, the factor 2 in the numerator accounts for the series connection required to obtain 24 V and the factors close to unity in the denominator represent a battery charging efficiency of 90%, a battery self-discharge level of 3% and a variability factor of 15% respectively.

The required battery capacity B in Ah is found from the relation[19]

$$B = \frac{731.6}{0.75} \times 4.2 \times \sum_j \frac{D_m - D}{D_m} \qquad (2.4)$$

where 731.6 is the average number of hours per month, 4.2 is the mean load current, 0.75 is the assumed permissible depth of battery discharge factor and the summation is over the months for which the daily average insolation D is less than the annual mean value D_m.

The values of A and B appropriate to the array inclinations given in Table 2.6 can be computed from Equations 2.3 and 2.4 and the results are   shown in Table 2.7.   As is to be expected the minimum array size increases as the tilt angle increases over the range given, but in a practical system a high tilt angle might be preferable as the associated battery capacity is less.   This is due to obtaining a more uniform distribution of collected radiation throughout the year.   The required battery capacity can also be reduced for any tilt angle by increasing the array size above the minimum value. Some results are shown in Table 2.7 and were computed by using the new values of A in Equation 2.3 to calculate effective values for D_m which were then used in Equation 2.4 to recalculate B.

The above calculations illustrate some of the technical considerations in the design of small photovoltaic power systems and can be used as a basis for the final choice of A and B, after taking into account the relative costs of arrays.

Table 2.6. Mean monthly values of insolation for Vancouver (latitude 49.2°N) for a South-facing plate inclined at various angles to the horizontal.[a]

TILT, DEG.	INSOLATION, $mWh\ cm^{-2}\ day^{-1}$												
	JAN.	FEB.	MAR.	APR.	MAY	JUN.	JULY	AUG.	SEP.	OCT.	NOV.	DEC.	MEAN
40	126	230	351	456	563	562	611	546	460	284	151	102	370
50	131	236	351	440	530	523	570	521	455	289	156	106	359
60	134	236	343	415	485	472	516	485	439	288	158	108	340

[a]Courtesy of J. E. Hay, University of British Columbia (data to be published).

Table 2.7. Array size and battery capacity to supply a 100 W, 24 V load in Vancouver.

TILT, DEG.	FOR CASE OF MIN. ARRAY SIZE		FOR CASES OF $A > A_{min}$.			
	A MODULES	B Ah	A MODULES	B Ah	A MODULES	B Ah
40	104	10807	108	10265	114	9660
50	108	10100	114	9357	120	8768
60	114	9350	120	8692	126	8114

2.6 GEOSYNCHRONOUS SATELLITE SOLAR POWER

The photovoltaic power systems that have been discussed thus far have all been located on Earth. An alternative site is in geosynchronous orbit and the concept of a satellite solar power station (SSPS) utilizing photovoltaic energy conversion was first presented in 1968.[70] Such a satellite would consist of a huge solar cell array maintained in geosynchronous orbit at 35,800 km from Earth together with dc–microwave converters and an antenna for beaming the microwave power to the ground, Fig. 2.16. Reconversion to dc electricity would be accomplished via a receiving antenna incorporating rectifying units, leading to an overall dc–dc conversion efficiency that could approach 70%.[71] An 18% photovoltaic conversion efficiency, to use a reasonable value for distant–future arrays, would thus be effectively reduced to 12.6% but this could be tolerated as the SSPS would be exposed to from 4 to 11 times the solar energy available to the sunniest areas on Earth.[72] In addition power would be generated throughout the day and continuously during the winter and summer peak demand seasons. An SSPS in geosynchronous orbit would be eclipsed only for periods up to 75 minutes around local–midnight during the three weeks before and after the equinoxes.[73] This would not be inconvenient as equinoxes are predictable and do not occur during times of peak energy consumption.

Electrical power levels at Earth between the limits of 2,000 and 15,000 MW, fixed by antenna geometry and satellite heat dissipation respectively, are possible and the 5,000 MW SSPS baseline configuration that has been adopted for feasibility studies in the United States is shown in Fig. 2.17. A 100 meter diameter central mast and dielectric support struts link the two solar cell array assemblies, between which is situated the microwave transmitting antenna. The solar panels are arranged to face the sun continuously and the antenna rotates twice a day to remain pointed towards Earth. Lightweight mirrors in the form of aluminized mylar or kapton are incorporated to provide a sunlight concentration ratio of 2 at the solar cell blankets. Solar cells that are both highly efficient and thin are required in order to minimize area and weight, and in addition must be encapsulated to guard against micrometeorite, electron and proton impact

Fig. 2.16. Concept of satellite solar power station. (After Glaser; courtesy of Amer. Inst. Astronaut. and Aeronaut.)

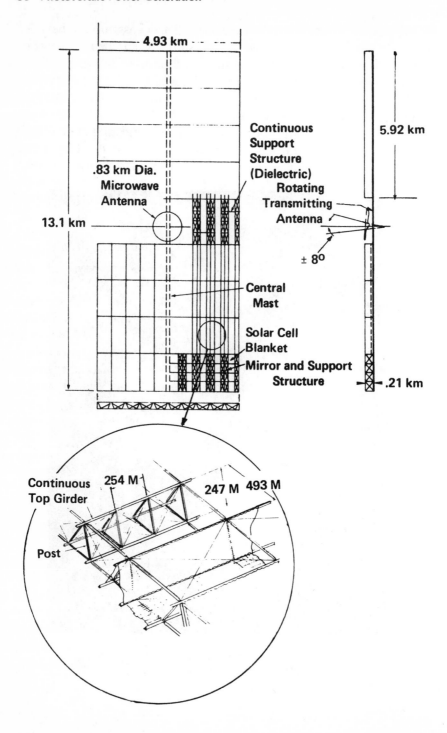

as well as harmful short wavelength radiation. The latter can be partially absorbed in selective coatings on the mirrors; micrometeorites are expected to cause an end-of-life cell loss of only 1%, but electron and proton damage could be severe and a logarithmic degradation in silicon cell performance leading to a 6% drop in efficiency over 5 years might be expected.[72] The enormous dc photocurrents (on the order of 10^5 A) would be arranged to circulate in the array structure in such a way that the induced magnetic fields would cancel out and so minimize satellite displacement by the Earth's magnetic field. However forces due to the Earth's gravitational field, solar pressure and microwave "recoil" would be present and would require compensation via thrusters situated on the array.[17]

Cross-field amplifiers called Amplitrons are being considered for the dc–microwave converters; a number of devices would be employed each operating at around 20 kV dc with a power output of 5 kW and conversion efficiency close to 90%.[71] As operation is in space, encapsulated vacuum enclosures need not, in principle, be used. The high conversion efficiency means that the use of passive heat radiators, possibly of pyrolitic graphite, can be contemplated. A series of generators would be combined in a subarray, about 5 meters square, forming part of the transmitting antenna. A closed loop, retrodirective-array phase-front control would be used with the subarrays to achieve the desired high efficiency, pointing accuracy and safety essential for the microwave beam operation.[72] An overall transmitting antenna diameter of about 1 km when combined with a receiving antenna about ten times this size would allow satellite to upper atmosphere transmission efficiencies to approach the theoretical limit of 100%.[71]

A 5000 MW SSPS would have considerable mass, even if the suggested power-to-weight ratio of 3.6 kg/kW was attained,[72] and so the problems of transportation and assembly are formidable. Perhaps the cheapest method would be to use a high thrust-to-mass shuttle for placement of SSPS components in low Earth orbit, where assembly would take place prior to geosynchronous orbit transfer using a low thrust-to-mass inter-orbit shuttle.[74] Onboard ion propulsion units, either nuclear or solar-powered, are being considered for the

Fig. 2.17. Satellite solar power station baseline configuration. (After Glaser;[72] courtesy of Amer. SES).

second stage transportation. For solar-powered units presumably the SSPS-generated electricity could be used, but a possible disadvantage to full solar array assembly at low earth orbit is the subsequent solar cell damage likely to be incurred on passage through the Van Allen belts. This might lead to the necessity of adopting a three-stage transportation process involving nonelectrical component assembly in low earth orbit, then transfer to an intermediate-altitude orbit above the Van Allen belts for array assembly and deployment prior to SSPS shipment to geosynchronous orbit.[17]   The common element in the two and three stage options is the first stage transportation and only through the use of reusable heavy-lift launch vehicles can the cost of freight delivered to low earth orbit be kept below $50/kg.[73] As each SSPS would require about 100 heavy-lift launch flights a dedicated shuttle system would be necessary and all this rocket activity might pose problems as regards atmospheric pollution (mainly water vapor and oxides of nitrogen) and sonic booms.   The space shuttle system currently under development in the United States may provide the necessary first step for SSPS technology verification and could be linked with the establishment of a space factory, engaging both human and tele-operators, for the space-based manufacture and assembly of SSPS components.

Assuming that a SSPS can be satisfactorily transported, assembled and located there remains the Earth-bound matter of collecting and harnessing the energy so generated.   The transfer of energy from space to ground has been assumed to utilize microwave radiation, as this has the desirable characteristics of high efficiency transmittance and conversion from and to dc electricity whilst employing reasonable antenna aperture sizes.   The incorporation of rectifying elements in the receiving antenna leads to the concept of a "rectenna" and this can be accomplished using a distributed array of half-wave dipole elements, integral low-pass filters, GaAs Schottky barrier diodes and bypass capacitors.[72]   A potential rectenna design[72] has the halfwave dipoles spaced about 0.6 wavelengths apart and arranged in a triangular lattice placed at a height from the ground of about 0.2 wavelengths, i.e., 2 cm if using 3 GHz radiation (which is attractive because of its low absorption by the Earth's atmosphere). The low-pass filter minimizes losses at the fundamental frequency and also assists in rejecting harmonics originating on rectification. Smoothing of the dc output is achieved with the by-pass capacitors each of which is positioned with respect to a diode so as to form a resonant circuit at the microwave frequency. The distributed nature of the system enables conversion efficiencies to be essentially independent of changes in beam direction and also of the variations in phase and power density that might be caused by nonuniform atmospheric conditions.

For a transmitting antenna diameter of 1 km, a 10 km diameter rectenna could collect 90% of the incident beam and could convert it to dc power at an efficiency of 85%,[72] thus no environmental problems as regards thermal pollution are anticipated. Public acceptance of hazards associated with the microwave radiation per se might not be so easily gained although no harmful biological effects seem likely. For example, birds or people in nonmetal aircraft flying through the main beam would experience some rise in temperature owing to absorption of microwave energy but this would not be expected to cause other than temporary discomfort as the maximum power density would not exceed 100 Wm^{-2}, i.e., the level used in the United States as the standard for long-term exposure. At the edge of the rectenna the power intensity would be two or three orders of magnitude less than at the center and not even a gross misdirection of the transmitted radiation should pose a threat to neighboring biological systems, because on the failure of the microwave beam-pointing system the microwave beam coherence would be lost and the beam would spread out so that power densities would not exceed those used in current communications systems.[72] In fact, it is radio interference with the latter systems that is likely to present one of the more serious side-effects of microwave usage. Since the microwave beam is not intended to transmit information it possesses no bandwidth for this purpose, however the electrical noise inherent to the system would be such that a 100 MHz-wide band would perhaps have to be dedicated to SSPS usage.[71]

It will be appreciated that the actualization of a satellite solar power station requires not only technological development in many areas but also social, political and international awareness and cooperation at a level much beyond that needed for totally Earth-bound photovoltaic systems. However the magnitude of the task is appropriate to the scale of disruption that would result if there was no proven system to generate baseline electricity when the present conventional fuels expire, be that from a mineral, ecological or political standpoint. As an interim measure it may be reasonable to test the concept of trunk lines of microwave energy by using such a medium for transferring large amounts of Earth-generated solar electricity from a sunshine-rich country to a more cloudy, electricity-hungry neighbor, e.g., from Australia to Japan or the United States. Appropriately placed satellites in geosynchronous orbit could provide the power-reflecting function,[75] and at the same time permit experience to be gained in the various space transportation and assembly techniques that will eventually be needed on a much larger scale for SSPS construction.

2.7 THE INSTITUTIONAL ASPECTS OF PHOTOVOLTAIC POWER DEVELOPMENT

The development and implementation of a major new technology in peacetime requires demonstration of technical feasibility, economic soundness and institutional compatibility. The institutional factor refers to the laws and customs of the land and is taken here to include the established practices and interests of industrial, commercial and governmental bodies. The technical and economic status of photovoltaics is the main subject of this book but some mention of the institutional barriers and incentives that are likely to affect the development of photovoltaics (and solar energy in general) is in order.

In the highly-developed countries of the free-enterprise world the relevant institutional factors are likely to include zoning and land use laws, rights to sunlight, building codes, patent rights, labor demarcation disputes, government subsidies to manufacturers and perhaps utilities, tax incentives to solar system purchasers, and the aesthetics and market value of new and retrofitted solar residences—in short, maintaining a certain standard of living in a capitalist environment. In underdeveloped countries the priorities are likely to be somewhat different and concerned with the improvement in living conditions, and hopefully with the avoidance of a two-tier (urban-rich, rural-poor) society. To achieve this requires an emphasis on rural development with attention being paid to upgrading the basic living environment, increasing agricultural productivity, and providing employment opportunities.[76] Much energy will need to be supplied to fulfill these ideals and, rather than embarking on a path of increasing dependence on depletable fossil fuels, it is sensible to consider the use of renewable energy sources for this purpose. For countries situated in much of Africa, Asia and Latin America the direct use of solar energy is the obvious choice, see Fig. 2.3. The distinct, yet often geographically-close nature of village communities in many of these areas would allow establishment of decentralized power plants particularly suited to solar fuelling. The use of photovoltaic-powered water pumps and educational TV systems has already been mentioned,[3, 10] and if this could be followed by water purification and household electrification then substantial improvements in the basic living environment could materialize. Further steps to provide solar energy systems for increasing agricultural productivity and then for industrial and community level uses would go a long way towards assisting in a beneficial development of Third World countries.[76] Photovoltaic power plants would thus seem to be an excellent commodity for use as aids to developing countries. The supply of such plants would also stimulate the solar cell manufacturing industry and assist in forcing solar cells down the "learning curve" towards lower prices.

The current high price of solar cells is undoubtedly one of the main reasons for the present lack of consumer interest in this product. The economic factors of photovoltaics seem to be the ones most often brought to the public's attentions, especially by large organizations with other energy interests, e.g., oil companies, some utilities and particularly the nuclear industry.[77] The facts that solar cells are currently following a 75% learning curve (unit costs dropping by 25% each time the production volume is doubled) and that the prospects of 5¢/kWh photovoltaic energy by 1986 is reasonable, are not generally known. The formation of national public interest organizations, dedicated to the advancement of photovoltaic (and other solar) technology,[77] would certainly assist in providing more accurate general information about photovoltaics and dispelling notions of its exotic and impracticable nature. Buildings and institutions with high public visibility, such as universities, schools, office blocks and shopping centers could play a major leadership role in solar energy education by incorporating model systems into their architecture.

The price of energy is, in most (if not all) instances, politically controlled and vociferous public support of photovoltaics would no doubt cause governments to provide some incentives for increased manufacture and purchase, so leading to a lowering of costs. At the national level such measures might include:[78-80] subsidies to solar cell manufacturers and perhaps utilities; tax credits to owners of solar houses, perhaps in the form of income tax deductions on mortgage payments; definition of solar energy apparatus to facilitate acceptance by labor unions, so obviating possible disputes about whether it is the job of an electrician, carpenter etc. to install photovoltaic systems; definition of maritime laws to assist in assessment of ocean-siting for large photovoltaic generating or receiving (from SSPS) stations; consideration of patent laws to guard against exclusive rights to use of major solar energy features, whilst at the same time maintaining the incentive to technological innovation that the present system allows. At the local government level incentives to implementation of solar energy systems might include: enlightened treatment of issues relating to zoning and land-use ordinances, building codes and materials, sun rights, community ownership of photovoltaic power plants; property and sales tax waivers; the required consideration of photovoltaics as a means to supply energy to new buildings and to existing buildings when matters of constructional additions or fuel system replacements arise. The demonstration of favorable government activity in the above areas would probably also stimulate financial institutions into investing capital into photovoltaics and establishing low interest loans, and advantageous insurance rates and warranties on solar conversion equipment.

Most of the above points apply to solar energy conversion systems in general, but a problem particularly germane to photovoltaics is the interface with the electric utilities. Some aspects of this problem were touched on in Section 2.3 and the main issue concerning proprietor-owned systems is whether the widespread use of photovoltaics will allow utilities to achieve any capital equipment and fuel displacement. If this cannot be done then the utility system load factor will decrease and owners of photovoltaic power plants can expect to pay more for their backup electricity. Monetary savings accruing to consumers due to the use of solar cells would then be proportionately lower than their actual energy savings from the utility.[81] Rate structures that are somehow equitable both to the utilities and to the small intermittent user of their services need to be worked out. By utilizing electricity storage and off-peak charging of that store, be it at the privately-owned point-of-use or at a utility-owned dispersed location, it should be possible to introduce proprietor-owned photovoltaic power systems in such a manner that they are compatible with utility economics and load management considerations (see Chapter 6). Utility-ownership of photovoltaic plant, irrespective of array location, would alleviate consumer/utility interface problems and allow photovoltaics to be viewed by the consumer as just another component in the electricity generation mix.

3.

Solar Cells: Basic Theory and Present Performance

Of all the various elements in photovoltaic power systems, the one currently being most intensively investigated is the opto-electric transducer itself, the solar cell. The various research and development efforts presently underway recognize that the widespread utilization of photovoltaic power depends to a large extent on the attainment of 12–20% efficient solar cells at much lower costs and higher production volumes than presently pertain. Solar cells and fabrication procedures that show promise of being able to meet these objectives are discussed in Chapters 4 and 5. In this chapter the basic factors affecting solar cell conversion efficiency are examined, and the performance of present solar cells is summarized.

Solar cells make use of the photovoltaic effect; namely, the absorption of photons to create equal numbers of positive and negative charges which can then be separated to develop a photovoltage and photocurrent, so allowing power to be delivered to an external load. The effect was first recorded in 1839 by Becquerel[82] who detected a photovoltage between AgCl and Pt electrodes immersed in an electrolyte. The first solid-state system known to exhibit the effect was based on selenium.[83] Further work in the 1930's on this material,[84] and on Cu-Cu$_2$O structures[85,86] pioneered the way to the successful development of exposure meters for photography, but little credence was given to the idea of utilizing the photovoltaic effect for producing more than miniscule amounts of power. It is

66

only since 1954, when the photovoltaic effect was demonstrated in silicon[87] that there has been a general realization of the ability of this effect to provide electrical energy on a scale, and in a manner, suitable, firstly, for spacecraft power supplies and, latterly, for significant uses on Earth.

The most suitable materials for solar cells are solid semiconductors in which an electrostatic inhomogeneity can be incorporated in order to separate the photogenerated electrons and holes. Both homojunction and heterojunction structures are possible, see Fig. 3.1. Homo-

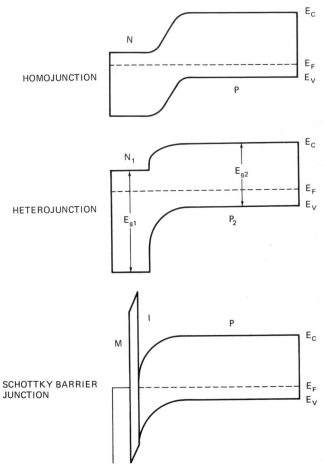

Fig. 3.1 Energy band diagrams of homojunction, heterojunction and Schottky barrier diodes.

junction solar cells are usually fabricated by conventional thermal diffusion techniques, although other methods under consideration are ion-implantation,[88] recoil-recombination[89] and epitaxy.[90] The latter technique is also suited to semiconductor–semiconductor heterojunction (henceforth called simply heterojunction) solar cell fabrication, with heteroepitaxial deposition being possible from both the liquid and vapor phases.   Schottky barrier solar cells are usually fabricated by vacuum deposition of a thin metal film onto the semiconductor, with or without trying to preserve a thin insulating layer at the interface.   Growth of the semiconductor on a metal substrate is another possibility.[91]

With the creation of a junction either in or on the semiconductor a rectifying structure results and a photocurrent can flow in the diode provided the photon absorption occurs within the vicinity of the junction region, i.e., within a distance equivalent to about one or two minority carrier diffusion lengths.   Electron-hole pair creation at greater distances from the junction is likely to result in bulk or surface recombination of the carriers, and thus a greatly reduced chance of carrier separation by the built-in field. The location of the photon absorption in the solar cell is governed by the absorption coefficient α of the material.   The magnitude of α depends on the physical structure of the material, on the densities of states in the conduction and valence bands, on the direct or indirect nature of the bandgap and on the photon wavelength.   The α values for some prominent semiconductor materials are shown in Fig. 3.2.   For Si which has an indirect bandgap the gradual change of α with wavelength means that photon absorption occurs from the surface to depths of several hundred microns.   For GaAs the steep absorption edge limits absorption to within about 2.5 μm of the illuminated surface.   50% of the AM1 spectrum is composed of photons with wavelengths less than 0.72 μm and thus is absorbed within the first 2–3 μm, even in the case of silicon.   It is clear then, that for efficient photocarrier separation, minority carrier diffusion lengths in silicon need to be considerably larger than in gallium arsenide, but that in both cases the electrostatic junction must be within a few microns of the irradiated surface.   Further requirements and characteristics of solar cells made from these and other materials are discussed in the following sections.

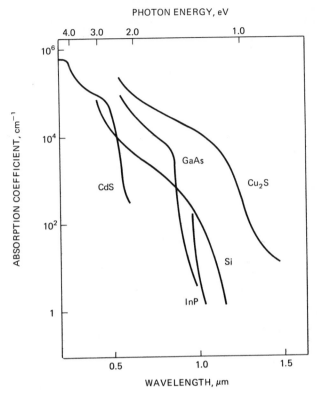

Fig. 3.2. Absorption characteristics of some solar cell semiconductors. (Adapted from Böer[22] and Wolf[92])

3.1 SOLAR CELL EQUIVALENT CIRCUIT

By connecting a load across the terminals of a solar cell a current I_L can flow through the load and develop a voltage V_L across it. The values of V_L and I_L, besides depending on the nature of the load, will be related to the photogenerated current I_P and the properties of the diode. These relationships can be established by reference to the simple equivalent circuit (Fig. 3.3) where imperfections in the diode leading to current leakage are represented by a shunt resistance R_{sh} and parasitic series resistance effects are represented by R_s. More exact equivalent circuits would take into account the distributed nature of both the current generator and the series resistance,[93] and also the

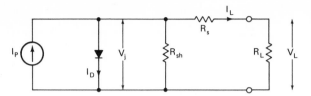

Fig. 3.3. Simple equivalent circuit of a solar cell.

fact that, because the diode is operating in photoconductance, the current components forming the normal diode dark current I_D and the light-generated current I_P cannot be adjudged to operate independently. However for many practical cases the principle of superposition, implied in Fig. 3.3, holds or can be used to develop suitable corrections,[94] and it follows that

$$I_L(1 + R_s/R_{sh}) = I_P - I_D(V_j) - V_L/R_{sh} \qquad (3.1)$$

where V_j is the voltage drop across the junction region of the diode due to either optical or electrical biassing. Equation 3.1 can be solved numerically in order to ascertain the effects of R_s and R_{sh} on the cell output characteristics,[95] but it can be appreciated that in general it is important for R_s to be small (particularly when using semiconductors with low bandgap (high I_P)), and for R_{sh} to be large (particularly in cases of high bandgap (high V_{oc})). Typical desired values for Si cells are $R_s < 0.5\ \Omega$ and $R_{sh} > 500\ \Omega$. The actual operating point on a given solar cell I–V characteristic is determined by the value of the load resistance R_L, and it is clearly desirable to choose R_L such that biassing occurs at the maximum power point I_m, V_m (Fig. 3.4). The energy conversion efficiency can then be described by

$$\eta = \frac{I_m \cdot V_m}{P_i \cdot a}$$

where P_i is the incident solar power density and a is the receiving area of the solar cell. The ratio of $(I_m V_m)$ to $(I_{sc} V_{oc})$ is defined as the fill factor, which enables a further expression for η to be written, namely

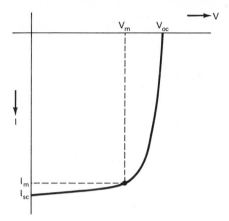

Fig. 3.4. Solar cell I-V characteristic.

$$\eta = \frac{FF \cdot I_{sc} \cdot V_{oc}}{P_i \cdot a} \qquad (3.2)$$

where I_{sc} is the maximum possible value of the photocurrent and V_{oc} is the maximum possible value of photovoltage. The active area of the solar cell will, in practice, be less than the full front surface area a because of the need to position opaque conductors on top of the cell to collect the generated current. Practical top contact structures are often in the form of comb-like grids so designed to optimize the trade-off between reduced active area and reduced series resistance brought about by increasing the contact coverage, Fig. 3.5. The grid structure improves R_s by reducing the lateral distance through the thin (and hence high sheet resistance) surface region that carriers must travel before collection. This surface layer resistivity would otherwise contribute significantly to R_s, the other components of which are semiconductor bulk resistance and contact resistance. Also shown in Fig. 3.5 is an antireflection coating and this is often employed in practical solar cells to improve the coupling of light into the semiconductor.

From Equation 3.2 it follows that for given insolation conditions and exposed surface area the conversion efficiency is related to three cell properties, namely short circuit photocurrent, open circuit

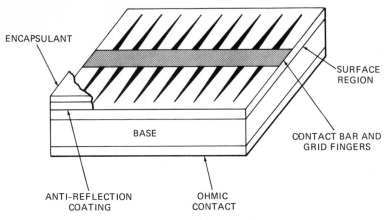

Fig. 3.5. Typical solar cell.

photovoltage and fill factor. The parameters affecting these three properties are discussed in the following sections, along with data on the magnitudes that have been realized in various types of practical cells.

3.2 THE SHORT CIRCUIT PHOTOCURRENT

A simple energy band diagram for the solar cell is shown in Fig. 3.6, where the base region has been taken to be a p-type semiconductor. With no illumination and the diode short circuited the net current flow is zero, since the saturation dark currents I_o, crossing the depletion region from each side of the junction, are necessarily equal. On illumination through the surface layer, electron-hole pairs will be created within the three regions of the cell, according to the absorption coefficients of the materials involved, and the generation rate can be expressed as

$$G(x) = \int_o^{\lambda_g} \alpha(\lambda)T(\lambda)M(\lambda) \exp\left[-\alpha(\lambda)x\right] d\lambda \qquad (3.3)$$

where $T(\lambda)$ is the transmittance into the surface layer, $M(\lambda)$ is the incident photon flux and λ_g is the wavelength of photons with energy equal to the bandgap energy E_g.

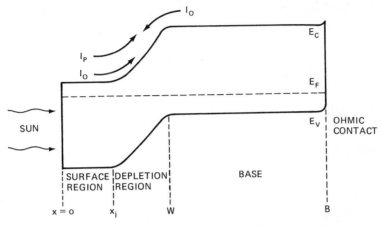

Fig. 3.6. General energy band diagram for a solar cell.

The concentration of excess minority carriers in the semiconductor regions of Fig. 3.6 can be computed by solving the appropriate continuity and current equations, namely

$$(1/q)(dJ_h/dx) - G_h(x) + U_h = 0 \qquad (3.4)$$

and

$$J_h = q\mu_h pE - qD_h(dp/dx) \qquad (3.5)$$

for the case of holes in n-type semiconductors, and

$$(1/q)(dJ_n/dx) + G_n(x) - U_n = 0 \qquad (3.6)$$

and

$$J_n = q\mu_n nE + qD_n(dn/dx) \qquad (3.7)$$

for the case of electrons in p-type semiconductors. In Equations 3.4–3.7, E is the electric field, U the rate of minority carrier recombination, p the excess hole density, n the excess electron density, μ the mobility, D the diffusion coefficient and q is the magnitude of the electronic charge.

Minority carriers generated within the field-free regions that diffuse to the depletion region will tend to be swept across the junction, as will be most of the carriers generated within the field region. In Fig. 3.6, electrons crossing from the right and holes crossing from the left become majority carriers and can be considered as leading to a photocurrent density which can be expressed as

$$J_P = J_s + J_d + J_b \qquad (3.8)$$

where the subscripts s, d and b refer to the currents originating from carriers generated in the surface, depletion and base layers respectively. Under short-circuit conditions no self-bias is generated by the separated carriers to reduce the external current and thus the latter assumes its maximum value, i.e.,

$$J_L = J_{sc} = J_P \qquad (3.9)$$

If certain high doping density and position-dependent composition effects are ignored,[96,307] Equations 3.3–3.9 can be used straightforwardly to calculate J_P for a variety of solar cell models. In many practical homojunction and heterojunction cells the condition of field-free regions depicted in Fig. 3.6, particularly in the surface layer, is not valid and the mobility and lifetime properties of the minority carriers must be given a doping density and field-dependence. The effect of the depletion layer field on minority carrier properties must also be taken into account, as must the various models for carrier recombination. Thus a complete solution of Equations 3.3–3.9 must involve numerical methods.[97-102] However, insight into the manner in which the various solar cell parameters affect J_{sc} can be obtained by calculating J_P assuming the simple model depicted in Fig. 3.6, and taking the case of low carrier injection levels to apply.

3.2.1 Homojunctions

To calculate the surface layer contribution to J_P in the case of an N/P homojunction, Equations 3.3–3.5 are first combined and the excess hole density profile computed by making use of the relevant boundary conditions, i.e.,

$$D_h \, dp/dx = S_F p \qquad \text{at } x = 0$$
$$p = 0 \qquad \text{at } x = x_j \qquad (3.10)$$

which imply that the minority carrier concentration at the surface is influenced by the surface recombination velocity S_F, whilst at the junction edge the minority carrier concentration is reduced to zero by the action of the depletion region field. The surface layer current density for a given wavelength then follows from Equation 3.5 evaluated at $x = x_j$ and can be written as[103]

$$J_s' = \frac{qTM\alpha L_h}{(\alpha^2 L_h^2 - 1)} \left[\frac{(H_h + \alpha L_h) - \exp(-\alpha x_j)(H_h \cosh Q_h + \sinh Q_h)}{H_h \sinh Q_h + \cosh Q_h} \cdots \right.$$

$$\left. \cdots - \alpha L_h \exp(-\alpha x_j) \right] \quad (3.11)$$

where L_h is the hole minority carrier diffusion length, $H_h = S_F L_h / D_h$ and $Q_h = x_j / L_h$.

In the depletion layer the electric field is sufficiently high that recombination of photogenerated carriers has little chance of occuring, and so the photocurrent density arising from photon absorption in this region of the cell is simply related to the number of photons absorbed. Thus, for a given wavelength

$$J_d' = qTM \exp(-\alpha x_j)[1 - \exp(-\alpha W)] \quad (3.12)$$

where W is the width of the depletion region.

To compute the photocurrent arising from photon absorption in the base layer the approach is similar to that used for J_s', with the boundary conditions in this case being given by

$$n = 0 \qquad \text{at} \quad x = x_j + W$$

and

$$S_B n = -D_n \, dn/dx \qquad \text{at} \quad x = B \quad (3.13)$$

The resulting photocurrent for a given wavelength due to photons absorbed in the base region is given by[103]

$$J_b' = \frac{qTM\alpha L_n}{(\alpha^2 L_n^2 - 1)} \exp[-\alpha(x_j + W)] \cdots$$

$$\left[\alpha L_n - \frac{H_n(\cosh Q_n - \exp(-\alpha B')) + \sinh Q_n + \alpha L_n \exp(-\alpha B')}{H_n \sinh Q_n + \cosh Q_n} \right]$$

$$\cdots (3.14)$$

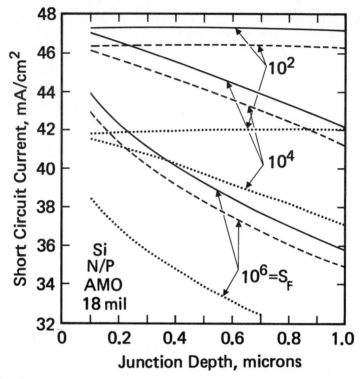

Fig. 3.7. Calculated AMO short-circuit photocurrents for Si N/P homojunction; solid lines: 10 ohm cm bases; dashed lines: 1 ohm cm bases; dotted lines: 0.1 ohm cm bases. Parameters of Table 3.1; $S_B = \infty$, $B = 450\,\mu m$. (After Hovel;[103] courtesy of Academic Press)

where L_n is the electron minority carrier diffusion length, $B' = B - (x_j + W)$, $Q_n = B'/L_n$ and $H_n = S_B L_n/D_n$, where S_B is the surface recombination velocity at the back of the cell.

The photocurrent density J'_P for monochromatic light is given by the sum of Equations 3.11, 3.12 and 3.14, and if a small wavelength increment $\Delta\lambda$ is considered the spectral current density can be written as

$$J_{P_s} = \sum_{\Delta\lambda} J'_P/\Delta\lambda \qquad (3.15)$$

and is usually expressed in units of $Acm^{-2}\mu m^{-1}$. The total photocurrent density J_P is then given by

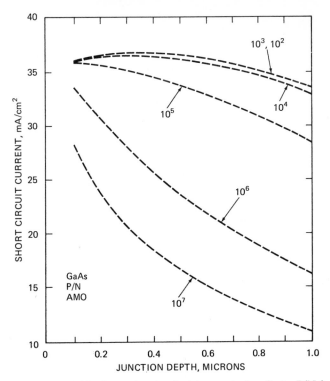

Fig. 3.8. Calculated AMO short-circuit photocurrents for GaAs P/N homojunction. Parameters of Table 3.1; $S_B = \infty$, $B = 15$ μm (Adapted from
Hovel[103] with permission).

$$J_P = \int_o^{\lambda_g} J_{P_s} \, d\lambda \qquad (3.16)$$

Values of J_P for typical Si and GaAs cells as calculated by Hovel[103]
are shown in Figs. 3.7 and 3.8, with the cell parameters being given
in Table 3.1. The results are shown for AMO insolation: under AM1
conditions the currents would be about 1/5 lower. For Si cells better
short circuit current performance can be obtained from higher base
resistivity cells. This is because of the sizable contribution to the
photocurrent of the base component J_b (due to the low α in Si).
J_b decreases at high doping densities due to the deterioration in
minority carrier properties, brought about by increased ionized
impurity scattering. The decrease of minority carrier diffusion

Table 3.1. Solar cell parameters for Si and GaAs homojunctions as used in Figs. 3.7 and 3.8.[a]

	N (cm^{-3})	D $(cm^2 sec^{-1})$	τ (μsec)	L (μm)	μ $(cm^2 V^{-1} sec^{-1})$	
Si N/P	5×10^{19}	1.29	0.4			surface
	1.25×10^{15}	36	15	232	1390 ⎫	base ⎰ 10 ohm cm
	1.5×10^{16}	27	10	164	1040 ⎬	⎱ 1 ohm cm
	5×10^{17}	10.9	2.5	52.2	420 ⎭	0.1 ohm cm
GaAs P/N	2×10^{19}	32.4	10^{-3}	1.8		surface
	2×10^{17}	5.7	1.58×10^{-2}	3.0		base

[a] From Ref. 103

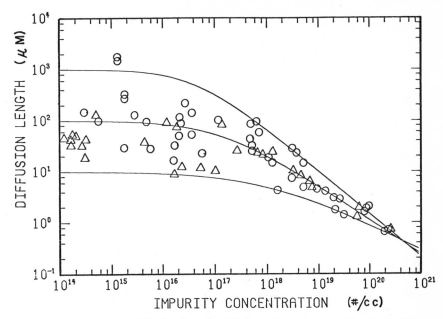

Fig. 3.9. Variation of diffusion length with impurity concentration in Si. Curves correspond to ranges typical of high, average and low lifetime. (After Hauser et al.;[98] courtesy of Pergamon Press)

length with doping density is shown in Fig. 3.9, and a similar relationship holds for the minority carrier lifetime.[105] Figures 3.7 and 3.8 also indicate the deleterious effect of a high value of S_F. The effect is more pronounced in GaAs then in Si because the higher absorption coefficient of the former ensures a larger contribution to J_{sc} from the surface layer photocurrent component J_s. It is this latter component that is affected by surface recombination, which serves to direct minority carriers in the surface layer away from the junction region.

3.2.2 Heterojunctions

Whilst the three regions of a solar cell as depicted in Fig. 3.6 are still relevant to the heterojunction case, some modification must be made to the energy band diagram to include the discontinuities in the band edges at the interface between the two component semiconductors.

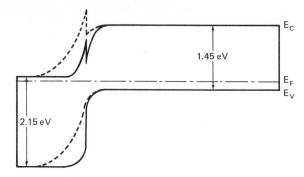

Fig. 3.10. Example of energy band diagram for an N/P heterojunction solar cell (*n*-AlAs/*p*-GaAs). Solid line, equal carrier concentrations either side of junction; broken line, *n*-side carrier concentration $\sim \frac{1}{3}$ that on *p*-side. (After Johnston et al.;[106] courtesy of Amer. Inst. Phys.)

Figure 3.10 shows the band diagram of one successful *N/P* heterojunction solar cell. Generally speaking band edge discontinuities resulting in "spikes" should be small in order that minority carrier flow across the junction not be excessively impeded. In the *n*-AlAs/*p*-GaAs example shown in Fig. 3.10 this is accomplished by having equal levels of doping density on both sides of the junction. Light is incident on the larger bandgap material, whilst the base region is formed by the smaller bandgap material and the depletion region has components within each of the two semiconductors.

The surface-layer contribution to the total short circuit photocurrent under monochromatic radiation is identical to that presented in Equation 3.11. Similarly the base-layer contribution is given by Equation 3.14, although in this case the depletion region must be identified as having components in both semiconductors, i.e., W_1 and W_2, and therefore the factor $[\exp(-\alpha(x_j + W))]$ in Equation 3.14 must be replaced by $[\exp - (\alpha_1 x_j + \alpha_1 W_1 + \alpha_2 W_2)]$, where α_1 and α_2 are the absorption coefficients in the "window" and base materials respectively. Because the depletion layer embraces the two semiconductors there are two components to the photocurrent contribution from the depletion layer, each having the form given by Equation 3.12. Equations equivalent to 3.11, 3.12 and 3.14 have been derived for *N/P* heterojunction structures by Milnes and Feucht[107] who also took into the account the small effect on the optical absorption of the refractive index difference between the two materials. The latter

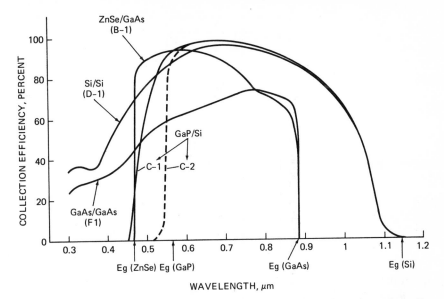

Fig. 3.11. Collection efficiencies of heterojunction and homojunction solar cells. Parameters of Table 3.2. (After Milnes et al.;[107] courtesy of Academic Press).

workers computed the spectral response to AM2 radiation of a number of heterojunctions that seemed suitable for solar cell use on account of their good lattice match and low values of energy band discontinuities. Some results, based on the data given in Table 3.2, are shown in Fig. 3.11 with the collection efficiency being defined as

$$\eta_{\text{coll}} = \frac{J_{P_s}}{qM} \tag{3.17}$$

Comparing the heterojunction and homojunction cells based on GaAs as shown in Fig. 3.11, it can be seen that the ZnSe–GaAs cell has superior response over much of the wavelength range. This is on account of the window effect which is possible with wide bandgap surface layers and which allows high energy photons to be absorbed at the junction region. In a homojunction cell these photons would be absorbed near the device surface, leading to a reduction in collection efficiency. The minority carrier properties of ZnSe are particularly poor, hence the sharp cut off at wavelengths corresponding to photon energies greater than the bandgap of ZnSe. The same phe-

Table 3.2. Solar cell parameters for various homo- and heterojunctions as used in Fig. 3.11.[a]

SOLAR CELL	x_j (μm)	B (μm)	S_F (cm sec^{-1})	L_n (μm)	L_h (μm)	D_n (cm^{2}sec^{-1})	D_h (cm^{2}sec^{-1})	N_D (cm^{-3})	N_A (cm^{-3})	τ_n (sec)	τ_h (sec)
n-p ZnSe-GaAs B-1	7.5	125	10^5	2.8	0.2	80	0.5	10^{18}	5×10^{17}	10^{-9}	10^{-9}
n-p GaP-Si C-1	0.5	500	10^5	215	0.3	23	1	10^{18}	2×10^{16}	2×10^{-5}	10^{-9}
n-p GaP-Si C-2	250.0	500	10^5	215	0.3	23	1	10^{18}	2×10^{16}	2×10^{-5}	10^{-9}
n-p Si-Si D-1	0.5	500	10^3	215	1.7	23	3	10^{19}	2×10^{16}	2×10^{-5}	10^{-8}
n-p GaAs-GaAs F-1	0.5	125	10^5	3.5	2	1.25	4	10^{18}	10^{16}	10^{-9}	10^{-8}

[a] From Ref. 107

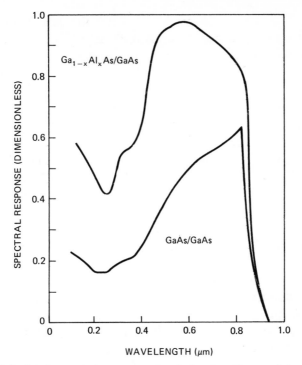

Fig. 3.12. Calculated spectral response for heterojunction and homojunction GaAs solar cells. In upper curve x in $Ga_{1-x}Al_xAs$ is 0.30 at front surface and is graded through the surface region to provide an aiding field. In lower curve $x = 0$. (Adapted from Hutchby et al.[109] with permission).

nomenon is noted in the GaP–Si cells, but some improvement accrues from reducing the GaP thickness, as exhibited by diode Cl in Fig. 3.11. Apparently by reducing this thickness to less than 2 μm the AM1 value of J_{sc} can theoretically exceed 30 mAcm^{-2},[108] but this is still less than the maximum value of 42 mAcm^{-2} appropriate to Si homojunction cells. The overall collection efficiencies of GaP–Si cells shown in Fig. 3.11 are less than in the homojunction case principally because, for silicon, the low absorption coefficient and gentle slope of the absorption edge mean that the base-layer response is little affected by the window properties of the surface layer. As far as photocurrent response is concerned then, heterojunction structures on Si would not be expected to yield any improvement over Si homojunctions. For GaAs, however, this is not the case and particularly promising window materials are AlAs[106] and $Ga_{1-x}Al_x As$.[109,110] In the latter case, by allowing the compositional factor x to vary with distance, a graded bandgap surface layer can result, leading to an improved minority carrier collection and substantial increase in response to the blue and green portions of the spectrum (Fig. 3.12).

Another heterojunction solar cell of interest is composed of Cu_2S/CdS, for which the energy band diagram is presently considered to be as in Fig. 3.13. In contrast to the heterojunction structures discussed above, the surface layer is the principal absorber in the usual mode of operation of this cell. The CdS forms a convenient base for the cells and its use has been maintained over the years, despite the poor lattice match between Cu_2S and CdS, because the Cu_2S layer can be formed from the CdS by a simple topotactic transformation. Minority carriers photogenerated in the Cu_2S diffuse to the junction where some recombination can be expected to occur via the large density of interface states ($\sim 10^{14}$ dangling bonds per cm^2.)[112] A further, much smaller, contribution to J_P results from the drift and diffusion to the junction of holes created in the CdS. There

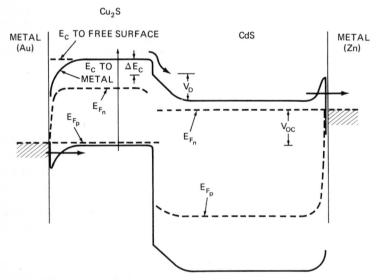

Fig. 3.13. Energy band diagram for Cu_2S/CdS solar cell under open-circuit illuminated conditions. (After Böer[111] with permission)

is presently some controversy over the factors limiting J_{sc} in Cu_2S/CdS cells. One suggestion,[111,112] is that high-field domains can form near the Cu_2S–CdS interface leading to field quenching conditions similar to those observed in photoconducting cadmium sulfide. The field quenching causes a transition from a high carrier lifetime state to a low carrier lifetime state and thus leads to a diode saturation current, which in this case would be the short circuit photocurrent, of somewhat lower value than that which would otherwise be supported by minority carrier extraction from the Cu_2S layer. In another model,[113] photocurrent limitations in Cu_2S/CdS solar cells have been attributed to more conventional mechanisms, such as photon reflection losses and photocarrier recombination at the interface, at grain boundaries, at the surfaces and in the bulk regions of the cell. The suggested maximum value of J_{sc} under AM1 conditions is 26 mA cm^{-2}.[113]

3.2.3 Schottky barriers

The generalization of the solar cell as a three-layered device as shown in Fig. 3.6 is still valid in this case, but the surface layer would be a thin, partially transparent metal. Neglecting the small barrier-lowering effect of the image force, the relevant energy band diagram for a metal/n-type semiconductor system is shown in Fig. 3.14, with ϕ_b being the barrier height. The surface layer photocurrent component can be calculated from Fowler's treatment of absorption in metals,[114] which leads to

$$J_s = \int_0^{\lambda_b} qM(1 - R(\lambda) - T(\lambda))\beta(\lambda)\,d\lambda \qquad (3.18)$$

where $R(\lambda)$ is the reflectance, λ_b is the wavelength corresponding to an energy equal to ϕ_b and β is the probability that an excited electron in the metal will be emitted over the barrier into the semiconductor. This simple treatment, which neglects electron–electron interactions in the metal and so avoids Monte Carlo calculations, is justified as J_s forms such a small part (typically $< 1\%$) of the total photocurrent.[101]

As can be seen from Fig. 3.14 the field region of the semiconductor extends to the metal surface and an inversion layer region will be

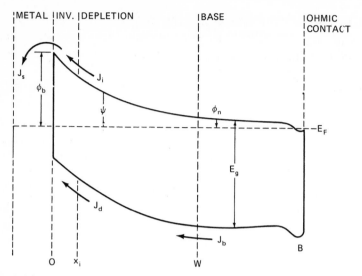

Fig. 3.14. Energy band diagram for metal/n-type semiconductor solar cell.

present for the case of high values of ϕ_b. The photocurrent component arising from absorption in the field region thus consists of two components, again of the form given by Equation 3.12. To establish the values for the depths of the inversion layer, x_i, and depletion region, W, which are needed in order to compute the field region photocurrents, involves computation of the voltage profile in the field region of the cell. This can be accomplished by solution of Poisson's equation

$$\frac{d^2\psi}{dx^2} = \frac{-q^2}{\epsilon} [N_D + N_V \exp(-(E_g + \psi)/kT)] \qquad (3.19)$$

where ϵ is the semiconductor permittivity, N_D the semiconductor doping density, N_V the effective density of states in the semiconductor valence band and ψ the potential energy parameter (Fig. 3.14). Equation 3.19 can be solved numerically, using the boundary conditions

$$\psi = -\phi_b \qquad \text{at} \qquad x = 0$$
$$\psi = -\phi_n \qquad \text{and} \qquad d\psi/dx = 0 \qquad \text{at} \qquad x = W \qquad (3.20)$$

From the resulting voltage profile the depletion layer boundary W

Table 3.3. Solar cell parameters for Au/Si and Au/GaAs Schottky Barriers as used in Figs. 3.15, 3.16 and 4.2.[a]

		Au thickness nm	ϕ_b (eV)	N_D (cm^{-3})	A^{**} (Acm$^{-2\circ}$K^{-2})	S_F (cm sec^{-1})	B (μm)
Si	n	10	0.9	10^{16}	110	10^5	250
	p				32		
GaAs	n	10	0.9	10^{16}	8.6	10^7	100
	p				79		

[a] For more details see Ref. 101.

can be found and interpolation can be used to find x_i, i.e., the point at which $\psi = -E_g/2$. The base-layer component of the short circuit photocurrent is of the same form as given for the homojunction case, Equation 3.14, taking into account the appropriate optical properties of the cell.[101]

To demonstrate the factors affecting J_{sc} as given by this model, consider the gold/n-type semiconductor system with parameters as in Table 3.3. The spectral current densities for Au/Si and Au/GaAs diodes are shown in Fig. 3.15. The dominance of the bulk component in the silicon case stems from the absorption coefficient properties of silicon. For GaAs, where α is both larger and shows a sharper edge, there is significantly more absorption in the regions of the cell close to the metal, and since quantum yields are high in this region the spectral current densities are also high. The widths of the inversion and depletion regions depend on the doping density (they are proportional to $N_D^{-1/2}$) and thus, as the doping density increases, a reduction in the contributions to the total short circuit photocurrent from J_i and J_d will occur. The doping density range available for Schottky barrier diodes runs from about 10^{14}–10^{17} atoms per cm^3. The lower limit is set by the bulk resistance of the device and the upper limit is governed by the onset of carriers tunneling through the depletion and inversion regions, leading to the loss of the rectifying nature of the contact. The effect of doping density on the photogenerated currents in Si is shown in Fig. 3.16. As described above the inversion and depletion layer photocurrents decrease with increase in doping density. The base component initially increases with doping density as the base/depletion interface

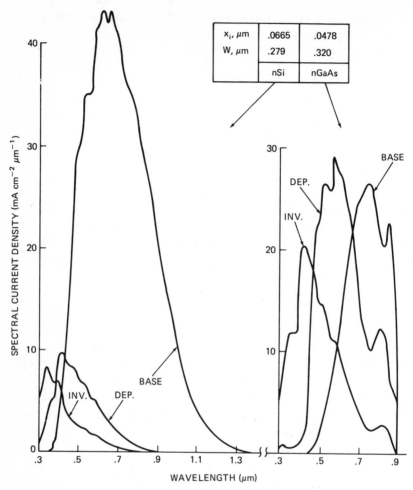

Fig. 3.15. Spectral current density for Au/n-Si and Au/n-GaAs Schottky barrier cells. Parameters of Table 3.3. (After McOuat et al.;[101] courtesy of Amer. Inst. Phys.)

moves closer to the metal-semiconductor interface i.e., the base region begins to encompass the region of high optical absorption and hence photocurrent generation. However, further increase in N_D (above about 2×10^{15} donors cm^{-3} in this case) leads to a deterioration in the minority carrier transport properties such that J_b begins to fall. The net result is a decrease in total photocurrent with doping

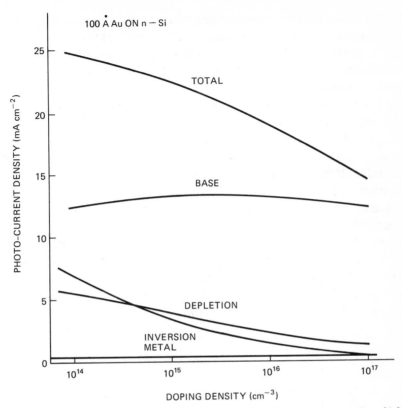

Fig. 3.16. Doping density dependence of J_{sc} for Au/n-Si solar cell.   (After McOuat et al.;[115] courtesy of IEEE)

density.   The dominance of J_b in Fig. 3.16 is not only a result of the absorption properties of Si but also attests to the long minority carrier diffusion lengths and lifetimes of this material.   In direct bandgap materials minority carrier lifetimes are small and thus the degradation in photocurrent with increase in doping density can be expected to be less pronounced than for silicon.

The Schottky barrier photocurrent decreases rapidly with increase in barrier metal thickness but is hardly affected by the value of ϕ_b, e.g., for Au on n-type (10^{15} donors cm^{-3}) Si J_{sc} decreases by about 40% on increasing the metal thickness from 5 to 20 nm, but changes only 0.4% on increasing ϕ_b from 0.6 to 1.0 eV.[101]

In the fabrication of metal-semiconductor diodes, achieving an intimate contact between the metal and the semiconductor has always been a problem. However, for solar cell operation, it has been realized that significant increases in performance can be achieved, either by preserving the very thin (0.5–2 nm) insulating layer that would naturally occur on the semiconductor prior to metal deposition, or by deliberately introducing such a layer. The resulting cell, termed MIS or surface insulator solar cell, is depicted in Fig. 3.17. The increase in performance of these cells (see Section 3.3.3) is not however related to an improvement in J_P; in fact, a deterioration in magnitude of the short circuit photocurrent can be expected if the insulating layer is allowed to become too thick (typically about 30 nm).[116,117]

In the Schottky barrier solar cell model presented here, the computed short circuit photocurrents are lower than would apply to homojunction cells fabricated from like semiconductor material. The reason for this is simply the loss of photons due to absorption in, and reflection at, the metal film. Antireflection coatings are

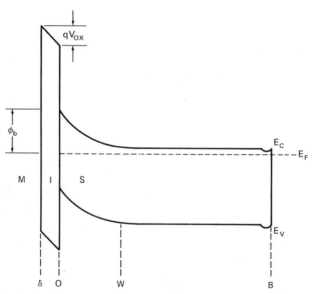

Fig. 3.17. Energy band diagram for metal/insulator/n-type semiconductor solar cell.

clearly a necessity for high performance Schottky barrier (MS and MIS) solar cells and their use should enable the improved short-wavelength response of these cells to be translated into an absolute improvement in total photocurrent.[300] By using antireflection coatings of refractive index greater than 2.1, reflection losses can be theoretically as low as 6% in Si and GaAs Schottky barrier cells.[118]

3.2.4 Measured values of short circuit photocurrent

The highest short circuit photocurrents that have been recorded to date are from silicon homojunction cells with textured surfaces, Ta_2O_5 antireflection coatings, shallow junctions and reduced doping densities in the surface layer. The last three properties, plus a very fine collecting grid pattern were features introduced in the "violet cell"[119] and led to improved J_{sc} values through an enhanced response to short wavelength radiation. Preferential etching of the silicon surface, using, for example, potassium hydroxide or hydrazine hydrate, can result in a light-trapping topography composed of pyramids of base size and height around 4–15 μm.[120,121] Combining this feature with the violet cell properties has led to a short circuit current equivalent to 36 mAcm^{-2} under AM1 conditions,[122] i.e., about 85% of the theoretical ultimate for Si. To date the most successful heterojunction cells using Si have employed wide bandgap window materials such as SnO_2 on n-Si[123] and $(In_2O_3)_{0.9}(SnO_2)_{0.1}$ on p-Si,[124] and yielded simulated AM1 J_{sc} values of 29 and 32 mAcm^{-2} respectively. These high values of photocurrent are probably only limited by series resistance and reflectance effects. Thus, even though the metal oxides can be deposited in such a way as to yield degenerate n-type materials with refractive indices close to 2 (at a wavelength of 0.5 μm), further improvement in J_{sc} will necessitate additional antireflection coating and contact processing. The use of TiO_2 as an antireflection coating has enabled AM1 J_{sc} values of 33 mAcm^{-2} to be attained in Schottky barrier (MIS) cells on silicon.[125]

For solar cells employing GaAs the short circuit photocurrents are higher in heterojunction cells than in homojunction cells. The main problems in the latter are high surface recombination velocity and the difficulty of shallow junction fabrication. The provision of a $Ga_{(1-x)}Al_x$As layer on top of a GaAs p–n junction can overcome the surface recombination problem and J_{sc} values of 28.5 mAcm^{-2} have been reported under AMl conditions.[126] A thin layer of $Ga_{(1-x)}Al_x As$ or AlAs directly on top of GaAs to form a conventional two-semiconductor heterojunction enables formation of shallow junctions, and AM1 J_{sc} values of 18 mAcm^{-2} for $Ga_{(1-x)}Al_x As$[127] and 28 mAcm^{-2} for AlAs[128] have been obtained. Further improvements with $Ga_{(1-x)}Al_x As$ can be anticipated on optimization of the graded bandgap properties of the surface layer. Heterojunction GaAs cells using SnO_2 surface layers have also been fabricated but present photocurrents are very low (11.4 mAcm^{-2} under AMO illumination)[129] probably due to high series resistance effects and poor interfacial properties. High short circuit photocurrents have been recorded in Au/GaAs Schottky barrier cells (26.5 mAcm^{-2} under simulated AMl conditions),[130] even without the use of antireflection coatings. This high value can be attributed to a very thin metal layer (6 nm) and a high minority carrier diffusion length (3.5 μm) in a 20–40 μm thick GaAs base layer grown epitaxially on a low resistivity substrate. The Schottky barrier structure is particularly suited to GaAs and related compounds as the high absorption coefficients ensure photocarrier generation in the regions of high field strength.

GaAs, Si and CdS represent the materials that have been most studied for solar cells use. Cells based on the latter material have utilized absorbing layers of Cu_2S,[113] CdTe,[131] $CuInSe_2$,[132] and InP.[132,133] The J_{sc} values under AMl illumination have reached 24.6, 19.3, 37 and 25 mAcm^{-2} respectively. The $CuInSe_2$ cells were only of very small area (to avoid microcracks) and the InP and $CuInSe_2$ cells utilized single crystal material, in contrast to the polycrystalline nature of the other structures. The lattice mismatches between InP and CdS and between $CuInSe_2$ and CdS are particularly low, 0.002 nm and 0.007 nm respectively, and the absence of any interfacial spikes in the conduction band edges enables attainment of good photoresponse. One proposal for improving the lattice match in Cu_2S/CdS cells is to dope the base layer with zinc to form a $Cd_{(1-x)}Zn_x S$ alloy. (The lattice constants of CdS, Cu_2S and ZnS are 0.4135, 0.396 and 0.3814 nm respectively.) Cells of this type have not yet realized improved J_{sc} values; this may be due to the presence of Zn in the Cu_2S layer, causing a reduction in minority carrier lifetime, or to the increased resistivity of $Cd_{(1-x)}Zn_x S$ layers, or to a spike in the conduction band.[135]

3.3   THE OPEN CIRCUIT PHOTOVOLTAGE

In addition to the photogenerated current, there are present in a solar cell one or more components of dark current. The magnitudes of the dark current components are related to the height of the electrostatic inhomogeneity at the junction, and increasing currents result on increasing the forward bias. Typical dark current transport mechanisms include injection of carriers over the junction barrier and subsequent diffusion, recombination of holes and electrons within the depletion region, and a combination of injection and recombination within the junction region assisted by tunneling via energy states within the energy bandgap of the semiconductor. The relative magnitudes of these components will depend upon the material and structure of the diode used. For example, in Si homojunction solar cells both injection-diffusion and depletion region recombination components are usually present; in Cu_2S/CdS heterojunction cells, tunneling through states at the interface between the two semiconductors is thought to be a major contributor to the dark current:[136] in $Al/SiO_x/p$Si MIS cells, for interfacial layer thicknesses up to about 2 nm, the dark current can be related to a tunneling component through the insulator as well as recombination within the depletion region and injection-diffusion into the semiconductor bulk.[137]

With no applied voltage or illumination, $J_P = 0$ and hence the saturation dark current densities J_o flowing in each direction across the junction must be equal. If the solar cell is now illuminated but the open circuit condition is maintained, again no net current can flow, but to balance the photocurrent component there must be an increase in the dark current which flows in the direction opposite to that of the photocurrent. This is achieved by a reduction in the potential energy barrier at the junction, with the result that the terminal voltage is in the direction of forward bias. Assuming superposition to apply,[94] the current balancing equation can be written as

$$J_P + \sum_D J_o = \sum_D J_o \exp (q V_j / \gamma k T) \qquad (3.21)$$

where the summations are carried out for the dark current components relevant to the particular solar cell, V_j is the voltage change at the junction brought about by the illumination and γ represents the factor that modifies V_j to an extent that will depend on the particular dark current transport mechanism. Under open circuit conditions

$V_j = V_{oc}$ and thus it is clear from Equation 3.21 that, to obtain a high value of open circuit photovoltage for a given photocurrent, the saturation values of the dark current components must be rendered small. The possible ways by which this might be brought about in homojunction, heterojunction and metal-semiconductor solar cells is examined from a theoretical standpoint in the next three sections, and upper bounds for the open circuit photovoltage are calculated for the three diode configurations.

3.3.1 Homojunctions

In homojunction solar cells fabricated from single-crystal semiconductor material tunneling, as a dark current mechanism, is unlikely to be important on account of the low doping densities used and the high degree of crystalline perfection at the junction. However the number of defects in the material increases with doping density and thus, when low resistivity devices are attempted, correspondingly more defects and associated energy states within the bandgap can be expected. This would lead to an increase in importance of the tunneling and recombination current components of the dark current, with tunneling starting to dominate, in Si for example, at base resistivities less than 0.01 ohm cm.[103] But in the usual Si and GaAs homojunction cells the tunneling component of the dark current can be ignored and thus in this section only the injection-diffusion and recombination currents are discussed.

The injection-diffusion current arises through the injection of majority carriers from either side of the junction into the oppositely-doped regions where the carriers assume a minority status and, for the case of no fields in the surface and bulk layers (Fig. 3.6), subsequently diffuse before recombining either within the semiconductor or at the surface. The behavior of the minority carriers is governed by the continuity and current equations (3.4–3.7) and these can be solved in the standard manner[138] to yield the injected current density, i.e.,

$$J_{\text{diff}} = J_{\text{od}} \left[\exp \left(q V_j / kT \right) - 1 \right] \tag{3.22}$$

where J_{od} is the injection-diffusion component of the saturation dark current density.

The space-charge region recombination current density J_{rec} arises through the carriers being injected from the neutral regions into the depletion layer and undergoing recombination therein without overcoming the full potential energy barrier at the junction. By considering recombination centers located near the center of the energy bandgap (i.e., the most effective carrier traps) and taking the electron and hole capture cross-sections to be the same, J_{rec} can be expressed as

$$J_{rec} = J_{or} \exp{(q V_j/2kT)} \qquad (3.23)$$

for values of V_j greater than a few kT/q.

Equations 3.22 and 3.23 indicate that injection-diffusion currents can be distinguished from recombination currents on the basis of the slope of a plot of $\ln J_D$ vs V, where J_D is the dark current density and V the applied forward bias voltage. The preexponential terms J_{od} and J_{or} are also dissimilar, leading to a situation where injection-diffusion will usually dominate at high values of forward bias and the recombination current, if present, will be manifest at low values of forward bias. This can be seen in Fig. 3.18 where the results of numerical calculations based on Si N^+/P diodes (see Table 3.4) are presented.[98] The dark current increases with base-layer resistivity and at the higher values is comprised almost exclusively of the injection-diffusion component. Increasing the base layer doping density reduces the base layer component of the dark current until eventually the back injection current (holes diffusing from the base to the surface layer) dominates. Thus further increases in base layer doping density do not reduce the total diffusion current, but do bring about a reduction in the base layer minority carrier lifetime. For very low base-layer minority carrier lifetimes the recombination current begins to become important as is evidenced by the approximate $\exp{(q V_j/2kT)}$ dependence shown in Fig. 3.18 for the lower base resistivities. The AMO values of open circuit photovoltage for solar cells with the dark current characteristics displayed in Fig. 3.18 are shown in Fig. 3.19. The higher V_{oc} values at higher base-layer doping densities are a direct consequence of the lower dark currents (Fig. 3.18 and Equation 3.21). However this capability of developing a high V_{oc} must be weighed against the reduced carrier lifetimes in

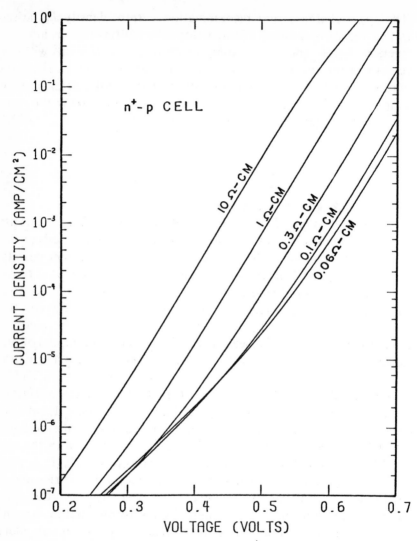

Fig. 3.18. Forward dark J-V characteristics of N^+/P Si solar cell for various base layer resistivities. Parameters of Table 3.4. (After Hauser et al.;[98] courtesy of Pergamon Press)

Table 3.4. Solar cell parameters for N^+/P and $N^+/P/P^+$
silicon homojunctions as used in Fig. 3.18.[a]

Overall cell thickness	250 μm
N^+ thickness	0.25 μm
P^+ thickness (N^+-P-P^+ cell)	0.5 μm
N^+ surface concentration	10^{20}/cm^3
P doping concentration	variable
P^+ doping concentration	10^{18}/cm^3
Lifetime in N^+ region	100 nsec
Lifetime in P and P^+ region	See Fig. 3.9
Surface recombination velocity	10^3 cm/sec
Antireflection layer	800 Å, SiO
Irradiance conditions	AMO

[a]From Ref. 98.

the base which will reduce J_{sc}. Thus the overall cell conversion efficiency will not increase indefinitely with increase in base-layer doping density. Furthermore voltage-reducing effects due to band-gap shrinkage and increased intrinsic carrier concentration effects also serve to reduce V_{oc} at high base doping levels.[139,140] The results of such effects are also shown in Fig. 3.19. The requirements of high open circuit photovoltage and high short circuit photocurrent are thus seen to be somewhat conflicting. However, high open circuit photovoltages are possible in solar cells with high resistivity base layers when an additional junction is incorporated into this latter region.[141] The resulting cell is termed a back surface field (BSF) device and the appropriate energy band diagram is shown in Fig. 3.20.

The BSF arrangement allows for the possibility of an increase in V_{oc} not only through a contribution to the photovoltage from the additional built-in barrier at the P/P^+ junction, but also through the "shielding" of minority carriers from the infinite surface recombination velocity of the back contact. This latter measure should increase J_P, provided diffusion lengths are long enough, by directing minority carriers that are photogenerated deep in the base layer towards the N^+/P junction, instead of allowing their recombination with photoexcited holes at the back contact. There is also a possibility of reduction in J_{diff} due to a reflection of injected minority carriers back to the N^+/P junction, i.e., an effective reduction in surface recombination velocity as "seen" by the electrons in the P-region cell. Some results for Si $N^+/P/P^+$ devices (Table 3.4) are

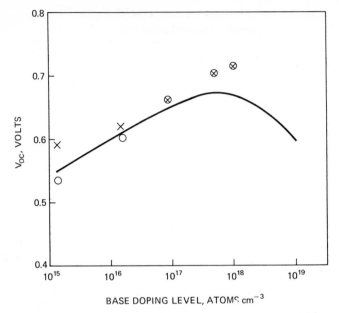

Fig. 3.19. Calculated AMO open-circuit photovoltages for Si N^+/P and $N^+/P/P^+$ cells. Solid line is theoretical curve allowing for high doping density effects. (After Godlewski et al.;[139] courtesy of IEEE); $0 \equiv N^+/P$ and $X \equiv N^+/P/P^+$ data. (Adapted from Hauser et al.[98] with permission).

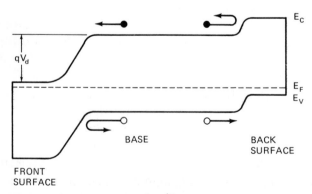

Fig. 3.20. Energy band diagram for $N^+/P/P^+$ back surface field homojunction solar cell. (Note reflection of minority carriers at the BSF.)

Table 3.5. Calculated values of V_{oc} for "well-matched"
heterojunction solar cells.[a]

		V_{oc}, V
n-p ZnSe-GaAs	B1	0.93
n-p GaP-Si	C1	0.65
n-p GaP-Si	C2	0.64
n-p Si-Si	D1	0.61
n-p GaAs-GaAs	F1	0.84
n-p $Ga_{(1-x)}Al_xAs$-GaAs		0.99

[a]From Refs. 107 (see Table 3.2) and 109.

shown in Fig. 3.19. For base resistivities less than about 0.3 Ω cm the reduction in height of the P/P^+ junction and the decreased minority carrier diffusion length prevent the BSF from enhancing V_{oc}. At smaller base thicknesses however, the beneficial effect of the BSF would extend to higher doping densities.

For GaAs homojunction diodes the low minority carrier lifetimes mean that J_{rec} dominates the total dark current to a much larger extent than in the silicon case.[142] Higher values of V_{oc} are possible with GaAs on account of its wider bandgap.

3.3.2 Heterojunctions

In addition to dark current components of injection-diffusion and recombination similar in form to those just discussed in homojunctions, a tunneling component is likely to be present in heterojunction diodes. Tunneling assumes importance in this structure because of interface states present at the metallurgical junction between the two semiconductors and energy states with the bandgap resulting from lattice mismatch, thermal expansion mismatch and also the effects of cross doping. Only in diodes with particularly good interfacial properties is the tunneling component likely to be negligible. Some such structures were discussed in Section 3.2.2 and the relevant V_{oc} values calculated on the basis of the models discussed therein are given in Table 3.5. It is evident that for both Si and GaAs, selection of a suitable semiconductor surface layer can, in theory at least, lead to significant increases in open circuit photovoltage over corresponding homojunction cells.

For heterojunction structures with poor interfacial properties tunneling components are likely to dominate and produce currents of the form[143]

$$J_{tun} = K_1 N_t \exp (K_2 q V_j/kT) \qquad (3.24)$$

where N_t is the density of energy states available for tunneling and K_1 and K_2 are constants. Equation 3.24 describes the case of tunneling via interfacial states but, in addition, mechanisms involving part tunneling and part thermal excitation across the junction can be present, thus contributing to the difficulty of predicting the performance of heterojunction solar cells based on semiconductor pairs of poor lattice mismatch. The Cu_2S–CdS system is a case in point and it appears that at room temperature not one, but two, independent conduction mechanisms involving tunneling-recombination processes can be present.[136] One of these mechanisms is probably associated with tunneling from a point part way up the barrier, thus adding a Boltzmann-type $\exp (K_3 q V_j/kT)$ term to the preexponential term in Equation 3.24. To further complicate matters thermal injection-diffusion currents arising from electron injection over the barrier into the CdS are also possible above room temperature.[144] Another important consideration in these cells is the fact that Cu_2S often penetrates down grain boundaries in the CdS leading to an effective junction area for the dark current that is much greater than the relevant area for photocurrent collection. For cells produced by techniques that would restrict this Cu_2S penetration (see Chapter 4), it is possible to predict that the likely maximum value for V_{oc} is around 0.54 V.[113] The use of Zn doping of the CdS to improve the electron affinity match could lead to V_{oc} values approaching 0.74 V.[113]

3.3.3 Schottky barriers

Even though the dark current in a metal-semiconductor diode can be described by an equation of similar form to those appropriate to the dark currents for homojunction and heterojunctions, the origin of the Schottky barrier current is fundamentally different from that of the injection-diffusion, recombination and tunneling components previously discussed. The forward bias dark current of Schottky barrier diodes, in the semiconductor doping level range of interest in solar cells, is determined mostly by the thermionic emission of majority carriers from the semiconductor into the metal.[145] The dark current can be represented by

$$J_D = A^{**}T^2 \exp\left(-q\phi_b/kT\right) \left[\exp\left(qV_j/mkT\right)\right] \qquad (3.25)$$

where A^{**} is the effective Richardson constant and m is the ideality factor which has a value very close to unity for intimate contact metal-semiconductor diodes. The presence of ϕ_b in the first exponential term of Equation 3.25 indicates the prime importance of this parameter in establishing the magnitude of the dark current, and thus V_{oc}. The ϕ_b is determined both by the difference between the metal work function and the semiconductor electron affinity and by the interface states at the metal-semiconductor junction.[138] For many semiconductors, including Si and GaAs, the surface state densities restrict the barrier heights attainable to values around $\frac{2}{3} E_g$ for n-type substrates. This means that saturation dark currents for metal-semiconductor diodes are generally higher than in homojunction diodes and thus V_{oc} values are small; e.g., the highest recorded barrier height for a metal on Si is 0.90 eV (Pt on n–Si), and a solar cell made from this combination could only be expected to yield an open-circuit voltage of 375 mV.[101]

The incorporation of a back surface field, which can improve the photovoltage response of homojunctions, is of little use in the MS Schottky barrier situation as the dark current is controlled by the front surface barrier, and the only improvement a BSF could bring about would be a slight increase in J_P. Any barrier that might be constructed to reduce the dark current must thus be situated at the metal-semiconductor interface. The MIS cell, introduced in Section 3.2.3, provides such a barrier in the form of a thin (1–2 nm) interfacial insulating layer. The insulating layer can either affect ϕ_b directly or impede the flow of majority carrier dark current, and each of these effects could serve to increase V_{oc}. When the presence of the interfacial layer is considered to change only ϕ_b, the dark current can still be represented by Equation 3.25, which implies that current control is through thermionic emission of carriers from the semiconductor to the metal and is not limited by transport through the interfacial layer. Under these circumstances calculations for silicon indicate that the barrier height can either decrease or increase with oxide thickness, for both n-type and p-type material, depending on the value of the metal work function ϕ_m.[146] High values of ϕ_m with n-type material and low values of ϕ_m with p-type material give high values of ϕ_b, but in both these instances the interfacial layer serves to reduce the barrier height. However, if charge is postulated to reside in the interfacial layer, then large increases in ϕ_b are predicted for the cases of negative charge with n-type material and positive charge with p-type material.[146]

In addition to the effect of barrier height modification the interfacial insulating layer can also serve to control the diode current and if the majority carrier current, which forms the dominant component of the dark current, could be reduced without any corresponding decrease in the minority carrier (photogenerated) current, then an increase in photovoltage response could be realized. Card and Yang,[116] for example, have shown that this is possible and can lead to a reduction in the usual Schottky barrier dark current (Equation 3.25) by a tunneling probability factor of the form $[\exp - ((4\pi/h)(2m^*\chi)^{1/2}\delta)]$, where h is Planck's constant, m^* the carrier effective mass, δ the oxide thickness and χ the tunneling barrier height. For this mode of operation the surface concentration of minority carriers is greater than the surface concentration of majority carriers, which means that interface states at the semiconductor-insulator boundary will be positively charged (for an n-type semiconductor) with respect to conditions in the dark. Under these conditions the contribution to V_{oc} from the voltage developed across the interfacial layer, V_{ox}, is negligible.[116] However, if a particular set of states localized at the semiconductor-insulator interface could be deliberately introduced with a majority carrier capture cross section much greater than that for minority carriers, it is, in theory, possible for the interface states to communicate mostly with majority carriers, thus decreasing V_{ox} and contributing further to the increase in V_{oc}. Fonash[147] has considered this possibility in detail, has termed the effect "field-shaping" and has predicted that open circuit voltages greater than the junction built-in potential are possible.

Yet another possibility exists for current transport control in MIS Schottky barriers and pertains to the case when the metal work function is chosen such that the semiconductor is inverted at the semiconductor-insulator interface over the voltage bias range of interest.[137,148] For an n-type semiconductor this means a high value of ϕ_m, and that the system fermi level at the semiconductor-insulator interface is nearer in energy to the valence band edge than to the conduction band edge. This can be appreciated from Fig. 3.17 when conditions of inversion prevail. The electron concentration at the oxide-semiconductor interface is much less than the hole concentration, and thus the majority carrier current is considerably reduced. There is no corresponding reduction in minority carrier transport and, indeed, under low forward bias conditions relevant to solar cell operation, large tunnel currents can flow, leading to a "pinning" of the minority carrier quasi-fermi level to the metal fermi level.[137, 148] The dark current thus becomes controlled by processes within the semiconductor and consists of injection-diffusion and recombination components appropriate to the induced p-n junction in the semiconductor. Under such conditions the dark current is given by Equations 3.22 and 3.23 rather than by the usual Schottky barrier Equation, 3.25, and much lower values are possible. Furthermore in this mode of operation the MIS cell can benefit from a BSF, so enabling further increase in V_{oc}.

3.3.4 Measured values of open circuit voltage

The maximum possible value of the open circuit photovoltage V_{oc} in a p-n junction is generally accepted[149] as being given by the junction built-in voltage, or diffusion potential V_d, see Fig. 3.20. At the illumination intensities appropriate to terrestrial sunlight V_{oc} values would not approach this magnitude, even under high concentration conditions. A high value of V_d necessitates high doping densities but in practical solar cells this leads to deleterious heavy doping effects (Section 3.3.1). For Si cells the practical maximum in V_{oc} is reached at a base resistivity of about 0.1 Ω cm and the highest recorded value (AM0 sunlight) is close to 635 mV.[150] In epitaxial cells with a graded base region (10^{16}–10^{18} donors cm^{-3}) a value of 636 mV has been obtained.[151] At these resistivity levels a back surface field is of no assistance in raising V_{oc}, but at resistivities of 10–100 Ω cm some advantage can be gained from such a structure and values of 600 mV have been reported.[141] Heterojunction solar cells using silicon as a base material have not led to any improvement in V_{oc} over the homojunction values. The highest reported values are 523 mV[123] and 510 mV[124] for cells employing conducting metal oxide surface layers. The use of p-Si with In_2O_3[124] and n-Si with SnO_2[123] minimizes the reduction in V_d due to poor electron affinity match, but problems of increased dark current conduction through interface states still exist and present a limit to further increases in V_{oc}. Some improvement might be expected from the incorporation of a very thin insulating layer at the junction interface, as has been found to be the case for silicon Schottky barrier cells for which a V_{oc} of 618 mV has been recorded under simulated AM1 conditions.[148]

In GaAs the prospect of higher photovoltages when employing heterojunctions rather than homojunctions has materialized. GaAs homojunction cells with $Ga_{(1-x)}Al_x$As windows have yielded V_{oc} values in the range 0.94–1.0 V,[126, 152, 153] and 0.95 V has been reported for a p-$Ga_{(1-x)}Al_x$As on n-GaAs cell under AM0 conditions.[154] Although V_{oc} values on GaAs Schottky barrier diodes are presently somewhat lower, the recently attained value of 758 mV on Au/oxide/GaAs cells represents a significant improvement over previous results.[155] Low values of saturation current (J_{os}) and high values of ideality factor (m) appear possible upon oxide formation on Ga-rich surfaces but a high J_{os} accompanies a high m for cells using oxides grown on As-rich surfaces.[155] Further increase in V_{oc} for GaAs MIS cells can be expected when these oxidation mechanisms are understood and antireflection coatings can be applied without degrading the photovoltage response.

For Cu_2S/CdS cells V_{oc} values are usually around 0.45–0.5 V and are limited principally by the electron affinity difference between the two semiconductors. Reduction of this mismatch by Zn-doping of the CdS appears to be possible and improvements in V_{oc} to around 0.68 V have been reported with $Cd_{(1-x)}Zn_x$S base cells.[135] Diffusion potentials around 1.1 eV are possible with cells employing CdS bases and either InP or CdTe absorbing layers, and this can lead to high values of V_{oc}, with 0.81 V[302] and around 0.73 V[157] respectively having already been obtained.

3.4 THE FILL FACTOR

3.4.1 Calculated values of fill factor

The fill factor of a solar cell is a measure of the "sharpness of the knee" in the output current-voltage curve (Fig. 3.4). It is clear that *FF* will be adversely affected by shunt and series resistances within the cell, but the dependencies on other properties of the cell may not be so obvious. To illustrate the situation by way of the derivation of an analytical expression for fill factor, it is convenient to make some simplifying assumptions, namely: to utilize the equivalent circuit of Fig. 3.3, to take $R_{sh} = \infty$, and to consider the diode dark current-voltage characteristic as being given by a single exponential relationship. The latter assumption still leaves the calculations with a measure of generality as dark current-voltage plots for some homojunction, heterojunction and most metal-semiconductor solar cells can be described this way (Section 3.3). The inclusion of R_s and not R_{sh} is justifiable for most small area solar cells fabricated on single crystal material; but may not be so either in large area cells on account of edge leakage and surface imperfection problems,[158] or in cells using polycrystalline material, where significant effects due to current leakage along grain boundaries and dislocations are possible.

With these assumptions and starting from Equation 3.1, the maximum output power can be found, from which it follows that

$$FF = \frac{I_m{}^2}{V_{oc} I_{sc}} \left[\frac{\gamma kT/q}{(I_P + I_o - I_m)} + R_s \right] \qquad (3.26)$$

where I_o is the saturation dark current and $\gamma = 1, 2$ or m for the situations described by Equations 3.22, 3.23 and 3.25 respectively.

From Equation 3.26 it can be seen that for given illumination conditions and cell temperature the parameters affecting the fill factor are series resistance R_s, saturation dark current I_o and the factor γ. To illustrate the situation Fig. 3.21 has been drawn for the case of a silicon solar cell at $300°K$, assuming that the generated (not necessarily the collected) photocurrent density is the maximum possible for AM1 sunlight (41.8 mA cm^{-2}). For a given value of I_o and for $R_s = 0$, the fill factor does not change with γ. This is because under these conditions the output voltage for a given output current increases monotonically with γ, so V_m and V_{oc} both change by a factor of γ, and as I_m and I_{sc} are unchanged, the fill factor (Equation 3.2) is constant. Increasing R_s for a given γ shortens the horizontal segment of the I–V curve (Fig. 3.4) yet doesn't affect V_{oc}, therefore FF decreases. But if R_s is kept constant ($\neq 0$) and γ is increased then both the horizontal segment and V_{oc} increase, and thus so does the fill factor. For a constant value of γ and for any value of R_s less than the resistive limit (i.e., that at which the I–V curve is a straight line be-

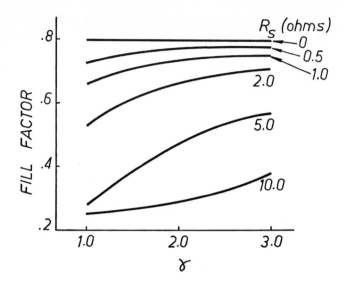

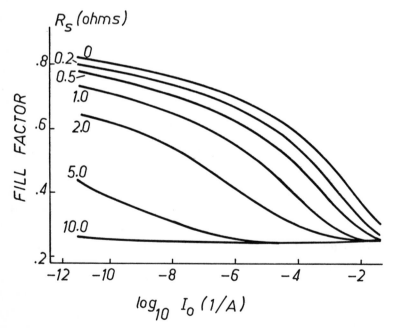

Fig. 3.21. Fill factor dependence on series resistance, diode exponential factor and saturation dark current.

tween I_{sc} and V_{oc}), increasing I_o decreases both V_{oc} and the horizontal segment of the I–V curve, and so the fill factor decreases rapidly (Fig. 3.21 b).

It can be concluded, that the fill factor is degraded by the presence of series resistance, can be improved (but not beyond the value for $R_s = 0$) by increasing γ, and is strongly reduced by increasing diode dark current. As, in practice, values of γ greater than unity are invariably associated with current transport mechanisms of high saturation current value, the fill factor under these circumstances will inevitably be less than in the case when, as in homojunctions for example, injection-diffusion dark currents dominate and $\gamma = 1$. This effect may not be apparent from earlier discussions of fill factor in which the interplay of γ and V_{oc} has been stressed and a reduction of FF with γ implied.[103,159] If γ is increased and V_{oc} determined to be constant, then this can only be achieved by increasing I_o and it is this latter increase, not that of γ that reduces the fill factor.

As can be seen from Section 3.3, in many instances the dark current voltage characteristic cannot be described solely by a single exponential relationship. Calculation of the fill factor under these circumstances has to proceed numerically as neither V_L nor I_L can be expressed solely in terms of the other. Wolf[160] has calculated that for a typical Si homojunction cell exhibiting both diffusion and recombination dark current components, the maximum output power is about 9% less than would obtain if only the diffusion current ($\gamma = 1$, low I_o) was present.

3.4.2 Measured values of fill factor

The requirements of low I_o and single exponential dark I–V relationship that would ensure a high value of FF, are most likely to be attained in either homojunction cells or heterojunction cells with good lattice matching properties. Furthermore, a low I_o is more likely to result from using high bandgap semiconductor materials. The required achievement of low R_s is almost entirely a contact technology problem. With these facts in mind it is not surprising that high fill factors have been recorded for GaAs homojunctions, 0.76–0.79,[103] $Ga_{(1-x)}Al_x As$–p–nGaAs diodes, 0.77–0.81,[103] and AlAs–GaAs devices, 0.82–0.85.[106] To reach fill factors of this level with Si has so far required the expedience of high technology processing, e.g., epitaxial growth of $P^+/P/N/N^+$ structures with a graded N-base region ($FF = 0.79$),[151] or $N^+/P/P^+$ diodes with intricate top contact metallization ($FF = 0.82$).[269]

For the present technology of glassy metal oxide–silicon heterojunction cells values of FF are in the range 0.64–0.70.[123,124] The introduction of a thin insulating layer between the metal oxide and semiconductor may improve the interfacial properties of this system and lead to higher values of FF. The interfacial insulating layer thickness is critical, however, and too large values can lead to photocurrent suppression through carrier recombination or series resistance effects.[161] Such phenomena have been detected in Al/I/pSi Schottky barrier cells where fill factors of 0.74 were obtained provided the insulator thickness was kept below 1.2–1.5 nm.[162] The use of an interfacial oxide in GaAs Schottky barrier cells has also allowed attainment of improved FF values, e.g., 0.78 for Au/I/GaAs[155] compared to 0.71 for Au/GaAs.[163] The highest reported fill factor for a Schottky barrier cell is 0.79 and pertains to the Au/GaAs$_{.78}$P$_{.22}$ system.[163] The attraction of this arrangement is the large semiconductor bandgap (1.69 eV) enabling attainment of high barrier heights (1.13 eV when using Au), and hence low values of I_o.

For Cu$_2$S/CdS solar cells the fill factor has long been in the range 0.65 to 0.70, but recent improvements in grid contact design and technology have increased the upper figure to 0.73.[113] Values around 0.76 and 0.66 have been obtained in CdS-based cells with InP[133] and CdTe[157] absorbing layers respectively.

3.5 THE EFFICIENCY

3.5.1 Calculated values of efficiency

The efficiency of a solar cell denotes the fraction of the incident illumination intensity that is converted to usable electric power, and as such is the number most appropriate to defining the quality of a cell. A convenient way of expressing the conversion efficiency is (Equation 3.2).

$$\eta = \frac{FF \cdot I_{sc} \cdot V_{oc}}{P_i \cdot a}$$

Thus for an efficient cell, high values of fill factor, short circuit photocurrent and open circuit photovoltage are required. It has been shown in this chapter that the factors affecting these parameters are of both a technological and a fundamental nature. One factor that serves as a ready descriptor of a semiconductor and can be used to make an initial selection of materials suited to solar cell use is the energy bandgap E_g. High values of I_{sc} can result from cells employing low bandgap semiconductors, whereas high values of V_{oc} and FF are possible with high bandgap semiconductors. Thus a plot of η versus E_g would be expected to show a maximum centered somewhere in the region of effective sunlight intensity (i.e., about 1–2 eV). It is not presently possible to obtain such a curve, other than by utilizing limit calculations, as the magnitudes of all the various parameters important in solar cell performance are not known for all the likely semiconductors. Figure 3.22 shows the results of limit calculations for homojunction cells subject to the following assumptions:[103] com-

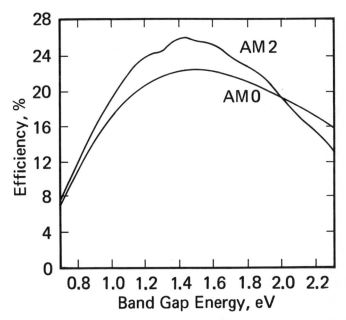

Fig. 3.22. Limit conversion efficiencies as a function of energy gap for AM0 ($P_i = 135$ mWcm^{-2}) and AM2 ($P_i \approx 74$ mWcm^{-2}). (After Hovel;[103] courtesy of Academic Press)

plete absorption of photons with energies greater than E_g; complete collection of all photoexcited carriers; ideal injection-diffusion dominated dark current with magnitude appropriate to 1 ohm cm Si; no series or shunt resistance effects; no contact area losses; no light reflection. Under these highly idealized conditions an optimum bandgap of around 1.4–1.6 eV is predicted, where the AM2 efficiency is around 25%. Similar results are obtained from limit calculations applied to Schottky barrier cells, subject to the same assumptions as above but with I_o calculated for limiting values of ϕ_b equal to E_g.[164] For heterojunction cells the wide range of possible semiconductor-semiconductor combinations precludes the carrying out of similar calculations, although some estimate of conversion efficiency can be made on the basis of the homojunction calculations for the material forming the lower bandgap semiconductor.[165, 166]

To compute more accurately the conversion efficiencies of particular solar cells, the more exact models referred to in this chapter must be used in conjunction with relevant data for the material properties. Silicon homojunction cells have been the most studied in this regard and recent calculations have focused on efficiencies in low resistivity base materials because of the potentially high conversion efficiencies brought about by increased photovoltage response.[97, 98, 105, 139, 167, 168] The cell features that would allow attainment of 20% AM0 (~23% AM2) conversion efficiencies are summarized in Fig. 3.23.[168] With respect to the model parameters listed in Table 3.4, the major improvements evinced by Fig. 3.23 are: optimizing the BSF by increasing the P^+ doping concentration and layer thickness by factors of 10 and reducing the P layer thickness to 100 μm; optimizing the base layer resistivity consistent with heavy doping effects and photovoltage enhancement (0.3 Ω cm was used for Fig. 3.23); incorporating an antireflection coating giving 5% reflection, i.e., a factor of 3 better than given by the arrangement in Table 3.4; obtaining high minority carrier lifetimes, particularly in the diffused N^+ layer. Not all these features are presently state-of-the-art but they do indicate the most likely directions research should take to maximize silicon conversion efficiencies.

GaAs, with a room temperature bandgap of 1.43 eV, is favorably situated in Fig. 3.22 and, from the earlier discussions on factors affecting I_{sc}, V_{oc} and FF, it is apparent that high efficiency cells require

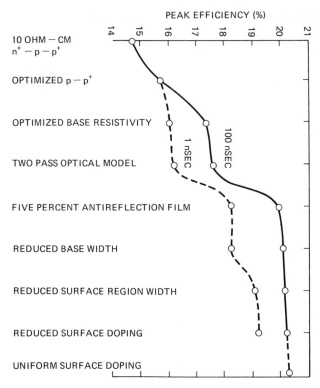

Fig. 3.23. Summary of calculated improvements in $N^+/P/P^+$ Si homojunction solar cells afforded by various geometry and doping modifications. See Table 3.4 for basic model parameters. (After Hauser et al.;[168] courtesy of IEEE)

some way of shielding a shallow junction from a high recombination velocity at the front surface. Heterojunction structures involving $Ga_{(1-x)}Al_xAs$ or AlAs offer the best way of achieving this. Detailed calculations pertaining to these structures,[99] and to $Ga_{(1-x)}Al_xAs/$ GaAs cells with graded bandgap surface layers,[109,110] indicate that AM2 efficiencies in the range 21–24% should be attainable. For Cu_2S/CdS cells the maximum possible AM1 efficiency appears to be about 10.4%,[113] and is determined principally by the interfacial properties of this combination. Improvement of these properties by Zn doping has already been mentioned and could lead to efficiencies approaching 15%.[113] The other approach would be to replace the Cu_2S

with materials having a better match to CdS; InP, CdTe and CuInSe$_2$ are possibilities and would seem capable of producing devices with efficiencies around 15%.

For Schottky barrier solar cells the principal impediment to attainment of high conversion efficiencies is the limitation imposed on V_{oc} by the saturation dark current. Direct bandgap materials with high values of E_g are particularly suited to Schottky barrier cells but, in the MS case, AM0 efficiencies in excess of 10% are unlikely.[101] However the incorporation of a thin interfacial insulating layer between metal and semiconductor affords a means of controlling I_o, such that efficiencies close to the values quoted above for homojunction and heterojunction cells should be attainable. Some calculations have been made for Al/SiO$_x$/p-Si structures and, even though silicon with its indirect bandgap and relatively low value of E_g is not conducive to extremely high performance, AM2 conversion efficiencies of 18.7% have been predicted.[169]

3.5.2   Measured values of efficiency

From the foregoing discussions it might be anticipated that solar cells based on GaAs would exhibit high values of conversion efficiency. This is indeed the case, although considerable technological development has been required to bring this about. The highest AM1 efficiency so far reported is 23% for a p-Ga$_{(1-x)}$Al$_x$As-p-GaAs-n-GaAs-n^+GaAs structure operating under conditions of concentrated sunlight.[170] This device utilizes liquid-phase epitaxial growth techniques which yield large minority carrier diffusion lengths; this factor and the provision of both back and front surface fields leads to very good carrier collection and low saturation dark currents. These features combine to provide the substantial increase in efficiency over the value of about 14%, which is the best reported for homojunction GaAs cells.[153]

The use of epitaxial processing has also been applied to Si devices and near-ideal cells with AM1 efficiencies of 12.6% have been reported.[151] This low value stems only from the use of a thin film structure (about 40 μm) which, in the case of Si, greatly reduces the photon absorption. Thicknesses down to about 50 μm might be consistent with high performance Si cells if textured surfaces are employed as, under these conditions, photon entry into the base layer is at nonnormal angles of incidence and thus higher path lengths for a given thickness can be attained.[122]   Such a structure has been employed in the Si cell for which the highest efficiency has been reported (19% AM1).[122] Reflection losses for this cell are only 3%, the fill factor is 0.77 and further improvement in performance can only be expected by increasing V_{oc}. The use of lower resistivity substrates (around 0.1 Ω cm) and reduced doping concentrations in the surface layer may achieve this end and an AM1 efficiency of around 22% appears possible.[171]   Other solar cell devices using silicon and showing reasonable efficiencies are MIS Schottky barrier cells (11.7% for Ti/SiO$_x$/p-Si)[125] and metal oxide heterojunctions (9.9% for SnO$_2$/n-Si[123] and ~12% for ITO/p-Si).[124] Improvements in V_{oc} and FF are required in these cells if significantly higher efficiencies are to be realized. Optimization of the oxide layer and antireflection coating technology in MIS cells and incorporation of an interfacial layer in the heterojunction cells may well bring this about. High values of V_{oc} and FF have already been obtained in Au/oxide/GaAs MIS cells, for which the highest reported efficiency is 15%.[130]

For solar cells utilizing CdS the highest efficiency obtained so far is 15% for a single crystal InP/CdS heterojunction.[303] For this combination the lattice match is excellent and the diffusion potential is high, but neither of these attributes are present in the much-studied Cu$_2$S/CdS system for which the best reported efficiency is 9.3%.[113] Zn-doping of the CdS base may improve this figure although present efficiencies in such cells are only 5%.[135] CdTe/CdS is another promising combination and recent success in the epitaxial growth of CdS on single crystals of CdTe has allowed attainment of an AM2 efficiency of 10.5%.[131]

4.

Solar Cells for Unconcentrated Sunlight Systems

To achieve techno-economic viability for use in large-scale electrical power generation, it has been estimated that solar cell arrays must be priced at about $0.50/peak watt.[11] In 1976 arrays of single crystal silicon cells could be purchased for $15/peak watt and, although this represents a halving in price since 1974, further reductions in the $/W index are clearly needed if the stated goal is to be met. High efficiency cells are likely to be very expensive and solar cells using very cheap materials and fabrication procedures are likely to exhibit low conversion efficiencies. Hopefully, in the intermediate efficiency range of 12–15%, there are solar cell materials suited to sufficiently low cost, mass-production, processing techniques that the $0.50/W goal can be attained. Efficiencies as high as 12–15% are probably necessary in order to keep costs due to land utilization and module interconnection at reasonable levels.[172] At the cell level, interconnection costs can be reduced by using large area solar cells, but not so large such that protection and voltage requirements (which demand considerable parallel and series connection of cells) cannot be met.

Conversion efficiencies in and above the 12–15% range have been attained in single crystal cells of Si, GaAs and InP/CdS, but the processing-technology currently required by these cells prohibits their cost-effective use in large-scale systems that do not utilize sunlight concentration. For these single crystal cells to be viable in the unconcentrated sunlight situation costs must be brought down by re-

114

ducing the expenses incurred in preparing the semiconductor material and processing the solar cell. Reducing the latter costs requires the utilization of "lower technology" procedures, whilst reducing material costs can be achieved by making less stringent demands on purity and by decreasing the volume of material required for a cell. As the collecting area of a cell must be large, the latter approach leads to the thin film solar cell concept, and this can be expected to be feasible for materials with high values of absorption coefficient. Crystalline silicon does not fall into this category and thus the simple thin film option is not open to this material.

A further possibility for reducing both material and processing costs would be available if the requirement of single crystallinity could be relaxed. A variety of methods are available for polycrystalline (and amorphous) semiconductor preparation and some of them may be compatible with large area, high throughput, solar cell fabrication techniques. Polycrystallinity raises questions of grain structure, grain boundaries and grain size. For solar cell applications it is clear that a fibrous or columnar structure is desirable, in which case each grain would behave similarly to a single crystal filament. Grain boundaries can be expected to provide internal surfaces with high surface recombination velocities, and thus to reduce their effect on photocurrent suppression they must be either passivated or reduced in number. The ability to achieve the latter is governed by the grain size. Larger grains are required in materials with low values of absorption coefficient and high values of minority carrier diffusion length. As silicon is the only semiconductor with low α and high L that is seriously being considered for solar cell applications, it follows that grain sizes in this material must be larger than in the other candidate materials.

Silicon is the most abundant, best understood and most technologically developed of semiconductor materials and for these reasons it is understandable that it should be considered for solar cell applications, even though its bandgap and absorption properties are not as desirable as some other semiconductors. There are encouraging indications that single crystal cell costs are amenable to lowering through the use of reduced-purity silicon and cheap production technologies, and that polycrystalline cells of the required grain size can be fabricated. Another cell combination that has been much studied for large area solar cell applications is thin film Cu_2S/CdS, but it is doubtful whether this cell can be expected to be viable for large-scale applications as its predicted maximum realizable efficiency is only about 10%.[113] However the use of Cu_2S/CdS cells in the smaller terrestrial systems outlined in Table 1.2, and currently monopolized by Si, is a distinct possibility. Other semiconductors, notably InP, CdTe and $CuInSe_2$, are suited to the forming of thin film hetero-junction cells with CdS and possess the capability of realizing the conversion efficiencies necessary for large power systems. Other semiconductor structures, particularly those based on GaAs, Cu_2O and organic semiconductors have also received some attention for unconcentrated sunlight, photovoltaic applications.

The state-of-the-art as regards low-cost, high-production-rate, solar cell fabrication procedures for the above materials is described in this chapter, following a brief discussion of some of the influences these procedures might have on properties important to conversion efficiency.

4.1 CONVERSION EFFICIENCY IN LARGE AREA SOLAR CELLS

The fundamental factors governing solar cell efficiency as discussed in Chapter 3 will, of course, apply to large area solar cells. However, because of the large area and low cost requirements some factors will be more affected than others and thus assume increased importance. Minority carrier properties and series and shunt resistance effects can be expected to be amongst these. In thin-film cells the semiconductor thickness may well be an important parameter governing the cell conversion efficiency, as would be the grain and grain boundary properties in polycrystalline cells.

4.1.1 Series and shunt resistance losses

The photocurrent, I_P, is directly proportional to the collector area and so in large cells it is particularly important to keep the series resistance at a low value. As the semiconductor surface and base layer resistivities are likely to be fixed by efficiency considerations, processes and designs that minimize contributions to R_s from contact resistance and collecting grid resistance must be sought. Low resistance contacts to semiconductors are more easily achieved if the material to be contacted is heavily doped or forms a metal/semiconductor interface with a large number of recombination centers. Whilst these conditions can be realized using conventional semiconductor technology, replacement of the latter with cheaper processes, e.g., the screen printing procedures used in thick film microelectronics technology, must be investigated. Some progress in this area has been reported.[173,174] The optimizing of front surface, metallization patterns for photocurrent collection involves a trade-off between the decrease in R_s and the decrease in active area brought about by having more contact fingers (Fig. 3.5). In principle the best configuration would be an infinite array of infinitesimally thin lines, but as practical line widths are finite the spacing between lines can influence the photocurrent collection and fill factor.[175-7] By replicating patterns a single large area cell can be treated as a number of smaller area units and if each unit has its own bond pad, then the burden of carrying a large photocurrent can be transferred to an external buss. Such an arrangement should allow the utilization of grid definition techniques having lower resolution and costs than photolithography, which is widely used at present.

The shunt resistance R_{sh} of a solar cell can be used to account for the effects of the various paths whereby current can by-pass the junction and lead to an increase in dark current (and hence a reduction in output voltage). The R_{sh} is thus associated with imperfections in the material, and an increase in its effect can be expected as the solar cell area is increased simply because of the longer perimeter and greater chances of incorporating areas of the surface containing defects.[158,178] Defects which allow diffusion piping are a particular problem in p-n cells which require shallow diffused junctions. In silicon MIS cells there is the possibility of edge leakage, due presumably to charge in the thin covering oxide forming an inversion layer over the whole (top and sides) surface. However, satisfactory passivation can be achieved by changing the surface potential of areas not directly under the barrier metal contact via a short heat treatment. [179,180] In polycrystalline solar cells some decrease in R_{sh} can be expected relative to the single crystal case on account of the shunting effects of grain boundaries. Despite the above possibilities for a decrease in R_{sh} it appears that large area solar cells can be made which are not usually limited by shunting effects. This is because increases in leakage current brought about by using reduced-perfection, larger area material can be overshadowed by the increase in photocurrent related to the larger area.

4.1.2 Minority carrier properties

Minority carrier lifetimes τ and diffusion lengths L are likely to become shorter as solar cell areas increase. This is because of the need to make large-area cells either from reduced-grade single crystal material or from thin-film material that is likely to be polycrystalline or relatively rich in defects. A reduction in τ and L affects the photogenerated carrier collection efficiency, and hence the short circuit current density J_{sc}, in all types of solar cell. In those structures where diffusion and recombination-type dark current components are possible, a reduction in τ and L can lead to increased dark

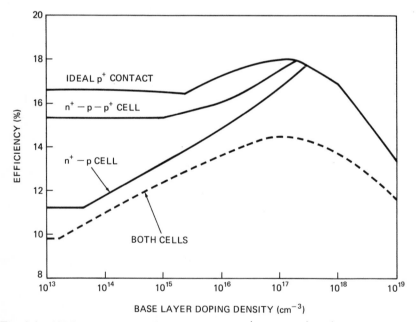

Fig. 4.1. AMO conversion efficiency for Si N^+/P and $N^+/P/P^+$ homojunction cells. Solid lines are for the high lifetime range shown in Fig. 3.9 and dashed lines are for the low lifetime range. Ideal P^+ contact has $S_B = 0$. Parameters of Table 3.4. (Adapted from Hauser et al.[98] with permission).

conduction, so adversely affecting V_{oc} and *FF*. The conversion efficiency is thus degraded, as illustrated by Fig. 4.1, which shows first-order calculations for Si N^+/P and $N^+/P/P^+$ cells based on the high and low ranges of minority carrier lifetime defined in Fig. 3.9.[98]

For Schottky barrier solar cells the dark current is usually governed by thermionic emission of majority carriers across the metal-semiconductor interface, and recombination-generation and diffusion currents only become important in minority carrier MIS diodes. Reducing τ and L in intimate contact Schottky diodes thus has its principal effect on J_{sc}. Conversion efficiencies for Au/Si and Au/GaAs cells are shown in Fig. 4.2, and were computed using the model presented in chapter 3.[101] The minority carrier lifetime has

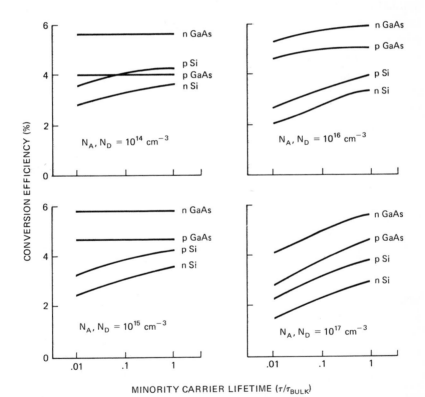

Fig. 4.2. AMO conversion efficiency for Au/Si and Au/GaAs Schottky barrier solar cells. Parameters of Table 3.3 with $S_F = 10^7$ cm sec^{-1}. τ_{bulk} data as a function of N_A, N_D taken from refs. 100, 181, 182. (After McOuat et al.;[101] courtesy of Amer. Inst. Phys.)

its largest effect on the base layer component of the photocurrent, thus (Fig. 3.15) Si cells are much more sensitive than GaAs cells to reductions in this parameter. The high absorption coefficient of GaAs keeps the response of this material high until the high doping densities greatly reduce the width of the surface field region.

4.1.3 Semiconductor thickness

The reduction in thickness of the absorbing part of a solar cell to values less than that of characteristic absorption lengths ($1/\alpha$) will clearly lead to a reduction in the number of photogenerated carriers. For silicon solar cells severe efficiency losses result at thicknesses less than about 100 μm, whereas cells with absorbing layers of, e.g., GaAs, InP, CdTe, Cu_2S, can, in principle, maintain near-bulk efficiency levels down to thicknesses of less than 5 μm. The use of textured surfaces, which ensure photon entry into the semiconductor at non-normal angles of incidence, might allow the use of silicon cells at thicknesses around 50 μm without there being any serious absorption penalty.[122] To give silicon a truly thin-film capability (efficient cells less than 10 μm thick) multiple-pass arrangements, in which photons are reflected off the back contact and then off the front surface etc., might be required.[183] However at such small thicknesses the nature of the back surface would start to have a serious effect on device properties. Shielding of the photocarriers from the infinite recombination velocity of an ohmic contact is clearly a necessity. In principle this is possible with a back surface field. Calculations for a perfect back surface field (S_{BSF} =0) are contrasted with data for an ohmic contact ($S_B = \infty$) in Fig. 4.3. In practice, though, the back surface recombination velocity is not zero and a reduction in base-layer thickness leads to an increase in surface recombination which could more than counteract any associated reduction in base layer recombination. Nevertheless a back surface field is necessary to maintain reasonable values of V_{oc} at low silicon thicknesses. Present results using epitaxially-grown layers confirm this with 594 mV having been obtained in 100 μm thick $N^+/P/P^+$ Si devices,[141] and 603 mV in 40 μm thick $N^+/N/P/P^+$ graded base Si cells.[151]

Whilst solar cells made from direct bandgap materials are more tolerant of a thickness reduction than Si cells are, there is likely to be a lower thickness limit (above that required by optical absorption

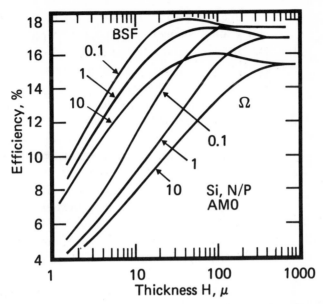

Fig. 4.3. Peak AMO conversion efficiency for Si N/P homojunction cells. Parameters of Table 3.1 with $S_F = 10^5$ cm sec^{-1}, $x_j = 0.2$ μm and aiding drift field in the surface layer, for BSF, $S_B = 0$, for Ω, $S_B = \infty$. (After Hovel;[103] courtesy of Academic Press)

considerations alone) determined by the resistivity, uniformity and integrity of the absorbing film. These factors, particularly the latter two, depend on the type of substrate used and the method of semiconductor film deposition. In thin film heterojunction solar cells it is possible that the properties of the wider bandgap material, rather than the absorbing layer, may dictate the overall cell thickness. The CdS/Cu$_2$S combination is a case in point. CdS films vacuum-deposited onto Ni–Fe substrates need to be about 10–30 μm thick to prevent subsequent Cu migration from the Cu$_2$S layer to the back contact of the cell.[172] However on preparation by pyrolitic spraying onto tin-oxide coated glass a dense yet transparent CdS film results allowing a thickness of 2–4 μm to be used.[184]

4.1.4 Grain properties in polycrystalline semiconductors

It is desirable that the method selected to grow polycrystalline solar cell material should yield fibrous or columnar grains. Otherwise the presence of grain boundaries parallel to the junction plane will lead to a reduction in effective cell thickness and a high value of effective back surface recombination velocity, both factors that could degrade cell performance if grain sizes were comparable to photocarrier diffusion lengths. In order to compute the expected conversion efficiencies from fibrously-oriented polycrystalline films, it is reasonable to model the grains as parallel cylinders separated by regions of high surface recombination velocity. Complete solution of the three dimensional diffusion equations with the eight boundary conditions appropriate to this model is a formidable exercise, further complicated by the need to consider an inhomogeneous spatial generation of photocarriers.[185] In the case of an intimate contact metal-semiconductor Schottky barrier diode the situation is simplified slightly by the fact that the dark current can be expressed solely in terms of the top junction parameters, and is not affected by grain boundary effects. Detailed calculations for this structure have recently been carried out,[102] and results for InP, GaAs and Si diodes are shown in Fig. 4.4.

As expected the materials with high absorption coefficients and low minority carrier diffusion lengths perform better than silicon, at least for small grain sizes. High resistivity films are favorable because of the increased width of the surface field region. A particularly important feature of Fig. 4.4 is evinced by the curve for 25 μm thick silicon which is still rising at the termination of the data points. If the film thickness was allowed to keep pace with the grain size, then the former's restrictive effect on the photocurrent would be progressively reduced. Less rigorous, but probably slightly pessimistic, calculations for polysilicon p-n junction cells not limited by cell thickness do not show signs of a conversion efficiency saturation with grain size even up to 1 mm grains.[186] The attainment of grain sizes of the order of millimeters in Si of thickness greater than about 200 μm should thus allow efficiencies around the 10% level. Silicon of this nature had recently been prepared by cutting slices from cast polycrystalline ingots and the resulting cells have achieved AM1 conversion efficiencies in the 10–14% range.[187,188] These devices were diffused p-n homojunctions, a fact which indicates that previous, widely-held beliefs that preferential diffusion of dopants down grain-boundaries would preclude this method of junction formation, were somewhat pessimistic.

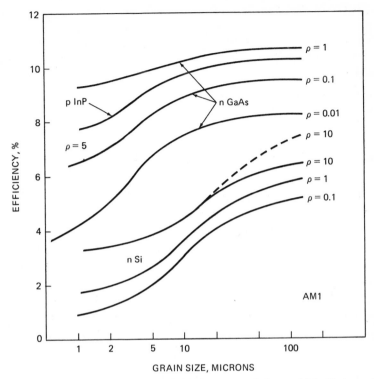

Fig. 4.4. Effect of grain size on AM1 efficiency of 2 μm thick GaAs and InP cells and 10 μm thick (solid lines) and 25 μm thick (dotted lines) Si cells, for various resistivities in ohm cm. ϕ_b taken as 0.9 eV and T(λ) as 100%. (After Hovel et al.[102] with permission).

The above results and calculations indicate that in polycrystalline material with fibrous grain structure the presence of grain boundaries need not be unduly detrimental to subsequent solar cell fabrication and performance. The influence of grain boundaries could probably be completely removed by doping them isotypically to form effective grain boundary surface field regions.[189,190] The practical danger here would be in causing a too-highly conducting boundary that might short-out the cell. For this reason it is encouraging that some practical polycrystalline cells do not appear to need this particular processing step.

4.2 SILICON SOLAR CELLS

Almost all the solar cells that have been hitherto used in terrestrial and space missions have been of the *p-n* junction silicon variety. Current commercial silicon solar cells are around 250 μm thick, utilize extremely high purity "semiconductor-grade" material and are fashioned from circular single-crystal wafers of 5–7.6 cm diameter. The cost of electrically interconnecting cells and then mechanically arranging them in modules and arrays is apparently as expensive as the solar cell fabrication itself, and about 30% of the latter expense is in the cost of the silicon material. The intensive efforts currently being devoted to establishing the viability of terrestrial photovoltaic power generation using silicon solar cells can thus be grouped into those seeking to define and prepare the grade of silicon that is really necessary for solar cells; those investigating means of growing single crystalline or polycrystalline silicon in a form suitable for large area use and mass production (about 5×10^8 m^2 of solar arrays annually),[92] and those studying various diode structures and cell fabrication techniques that might be compatible with the type of silicon resulting from the above processes. The present activities in these three areas are reviewed in the following sections.

4.2.1 Solar-grade silicon

The present route followed by silicon on its way from quartzite in the ground to a single crystal solar cell can be summarized as[191]

$$SiO_2 \rightarrow MG\text{-}Si \rightarrow SiHCl_3 \rightarrow Poly\ SeG\text{-}Si \rightarrow Single\ SeG\text{-}Si \rightarrow Wafer \rightarrow Cell,$$

where MG and SeG refer to metallurgical and semiconductor grade respectively. MG-Si at a cost of around \$1/kg is produced in submerged-electrode arc furnaces by the carbon reduction of quartzite. The molten silicon is tapped from the furnace and blown with oxygen or an oxygen-chlorine mixture to reduce the content of Al, Ca and Mg. After solidification and pulverization $SiHCl_3$ is manufactured by fluidizing a bed of fine MG-Si particles with hydrogen chloride in the presence of a copper-containing catalyst. The trichlorosilane is then converted to polycrystalline SeG-Si by chemical vapor deposition onto heated silicon substrates. The polycrystalline SeG-Si, can be purchased at about \$65/kg, with about 90% of this cost attributable to the processes involving $SiHCl_3$. To attain solar-grade Si,

SoG-Si, at the level of $10/kg (the goal set in the United States by ERDA) the $SiHCl_3$ purification step must be circumvented. Ways in which this might be achieved can be selected once the impurity levels that can be tolerated in SoG-Si have been defined. Some progress in this direction has been made by intentionally doping n- and p-type SeG-Si with secondary impurities known to be present in MG-Si and then observing the pertinent electrical properties of solar cells fabricated in the conventional manner.[192-4] Interpretation of the results is not straightforward as interactive effects between various secondary impurities could occur, nevertheless the summary presented in Table 4.1 gives some indication of tolerable secondary impurity levels. More attention has been given to p-type material owing to the predominant use of N^+/P Si homojunction cells. Tolerance of solar cell material to Ti, Zr, V and possibly Na is particularly poor, but most other impurities appear to be tolerable at the 10^{15} cm^{-3} or higher level. There is even some evidence that small amounts of Ni, Mn, Mg, Cr and C may actually lead to some improvement in solar cell performance.[193]

Two approaches to producing SoG-Si without using trichlorosilane and which show promise of being able to meet the $10/kg cost goal,

Table 4.1. Permissible Secondary Impurity Concentrations in 0.5 Ω cm Si such that a 10% Efficient Solar Cell can be Fabricated from it.[a]

SECONDARY IMPURITY	PERMISSIBLE CONCENTRATION	
	B DOPED	P DOPED
Ti	4×10^{13}/cc	3×10^{14}/cc
Zr	4×10^{13}/cc	---
V	1×10^{14}/cc	---
(Na)	2×10^{14}/cc	---
Cu	1×10^{15}/cc	---
Fe	1×10^{15}/cc	---
Mg	$>2 \times 10^{15}$/cc	---
Cr	5×10^{15}/cc	---
Mn	2×10^{16}/cc	---
Ni	$>4 \times 10^{16}$/cc	---
B	$0.8(Cz); 10.(Fz) \times 10^{16}$/cc	---
C	$2.0(Cz); 0.8(Fz) \times 10^{17}$/cc	5×10^{17}/cc
Al	3×10^{17}/cc	---
P	---	8×10^{16}/cc

[a]From Ref. 193. Permissible concentrations are uncertain by about one order of magnitude.

are the Zn reduction of $SiCl_4$ in a fluidized bed,[195] and the use of improved raw materials in the usual arc furnace reduction process followed by either blowing reactive gases (Cl_2, O_2, HCl) through molten MG-Si or acid-leaching crushed blocks of polycrystalline MG-Si.[196] If the SoG-Si is to be used as stock for single crystal growth then the crystal-growing step can provide further material purification, e.g., starting from regular MG-Si material, 2% efficient cells have been made from second generation Czochralski ingots.[197] By employing slow growth rates all transition element concentrations can apparently be reduced below 50 ppba, and subsequent float-zone refining has already produced materials from which solar cells with an AM0 efficiency of 10.7% have been fabricated.[196]

4.2.2 Silicon wafer preparation

SeG-Si crystals grown by the Czochralski technique provide over 90% of the single-crystal wafers utilized by the semiconductor industries of the world.[198] This large market has resulted in a very well-defined product, with crystal diameters of 7.6 cm now being standard and increases of up to 10 or 12 cm being investigated. However, as regards being able to supply the projected production rate for solar arrays (5×10^8 m^2 yr^{-1}),[92] the Czochralski technique is limited not only because it produces circular wafers of relatively low packing density but also on account of the nature of the necessary wafering process.

Present slicing of silicon ingots usually involves single inner-diameter blade saws which can produce approximately 25 7.6 cm wafers per hour with a kerf loss for 0.025 cm thick slices of about 0.03 cm.[198] The process is thus not only slow but also wasteful of material. Other estimates[191] suggest that the silicon yield going from single crystal SeG-boules to 5 cm wafers is only 20%, and that the overall silicon yield from SiO_2 in the ground to finished solar cell is only 3.2%.

Fig. 4.5. Heat exchanger method of crystal growth. (i) Experimental arrangement (After Schmid [200]; courtesy of McGraw-Hill); (ii) Schematic drawing of growth process: (a) prior to melting; (b) starting material melted; seed partially melted to ensure good nucleation; (c) liquid-solid interface expands in nearly ellipsoidal fashion; (d) crystal growth completed (After Zoutendyk [201] with permission).

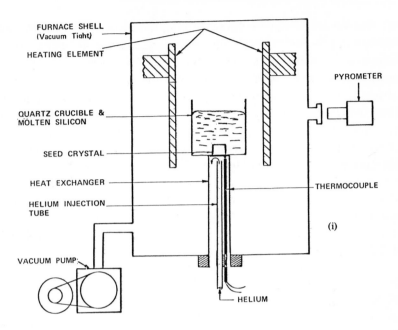

(i)

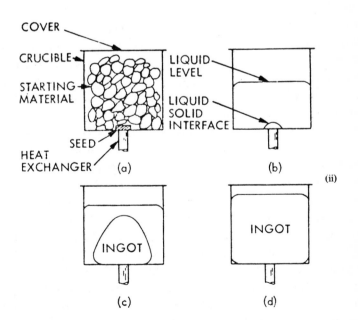

(ii)

Clearly there is room for improvement in matters of silicon recovery during processing. As regards boule cutting speed, some increase might be possible as solar cell wafers could be less sensitive to surface damage than are slices used for other electronic devices. But any gain in this regard is likely to be offset by the use of larger diameter ingots, which would require slower cutting speeds. Multiple-blade or -wire cutters operating in an abrasive slurry medium are capable of slicing more than one crystal at a time and rates of 48 to 90 slices per hour have been predicted for 7.6 cm diameter wafers.[198] To extrapolate these figures to likely values for 10 or 12 cm wafers is speculative, but if it is assumed that two 12 cm crystals can be sawed simultaneously at the experimentally observed rates for 7.6 cm crystals then Czochralski wafer costs can be projected to become as low as $30–36/m^2.[198] This figure assumes the use of $10/kg poly SoG-Si, a pull rate of 15 cm hr^{-1} (theoretical maximum for a 12 kg charge is 18 cm hr^{-1})[198] and realistic capital costs, operating schedules and depreciation times. The estimates of Wolf[92] fall in a similar range and represent an order of magnitude cost reduction over that of present silicon substrates; e.g., at $33/m^2 for a 12 cm diameter slice the material cost per peak kilowatt is about $330, as opposed to the current figure of about $3,500/kw.[188]

Another silicon ingot growth method that shows particular promise for photovoltaic applications utilizes directional solidification in conjunction with a novel helium heat-exchanger arrangement.[199,200] A typical setup is shown in Fig. 4.5: the seed bonds to the crucible bottom during the heat-up cycle and partial melt-back of the seed occurs. Growth is controlled by increasing the helium flow rate (to decrease the solid temperature gradient) whilst simultaneously decreasing the liquid temperature gradient by lowering the furnace temperature. With no moving parts, vibration and turbulence effects are absent, the operation under vacuum removes volatile impurities, and any contaminants from the crucible float to the surface and are not incorporated in the crystal that grows upwards and outwards from the bottom. This three-dimensional growth leads to high growth rates and, as the final crystal assumes the shape of the crucible, suitable configurations such as squares or hexagons can result. Present problems include cracking on final solidification, carbon contamination from the graphite resistance furnace and the possibility of very slow cutting speeds for the projected ingot diameters of 15 cm.[200] Some improvement in the slicing process should result from the use of multiple-sawing techniques, perhaps using diamond-impregnated, copper-coated steel wires.[199]

4.2.3   Silicon sheet preparation

If silicon could be pulled from the melt in sheet form then the time-consuming and costly process of wafering could be avoided altogether.   Two techniques whereby this might be achieved can be classed as self-shaping growth from a melt in a crucible, or growth using shaping dies which are in direct contact with molten silicon in a crucible.   Dendritic-web growth falls into the former category and the growth geometry for the process is shown in Fig. 4.6a. The web portion of the crystal is initially a film of liquid bounded by the button and surrounding dendrites.   As this film freezes the web and dendrites form an essentially single-crystal sheet, with the only flaw being one or more twinned lamellae running through the center of the web and parallel to the surface.   Solar cells made from this material do not seem to be adversely affected by this structure and cells recently processed on material made in the 1960's show comparable performance to regular Czochralski cells.[202]   Successful dendritic web growth depends very largely on maintaining the required temperature gradients within the melt.   To achieve this a new thermal design of crucible has recently been proposed with a temperature profile compatible with the thermal field of the web itself,[202] and with this arrangement the operational goal is for a semi-automated, quasi-continuous growth (120 hours) yielding 50 mm wide dendritic web of 0.1–0.22 mm thickness.   No cost analysis for this process is presently available, but certainly the end product is in a very desirable form for terrestrial solar cell arrays.   This is also the case for another self-shaping growth method, called the lateral pulling method, in which a trough-shaped crucible is kept full of molten silicon so allowing, after seeding, a solidified silicon sheet to be withdrawn almost horizontally.   High growth rates are possible with this method owing to the large area of the solid-liquid interface and the high rate of heat dissipation permitted by the growth arrangement.   Single crystal silicon ribbon of 1–5 cm width and quality close to that of dislocation-free Czochralski material has already been produced at a growth rate of about 40 cm/min.[304]

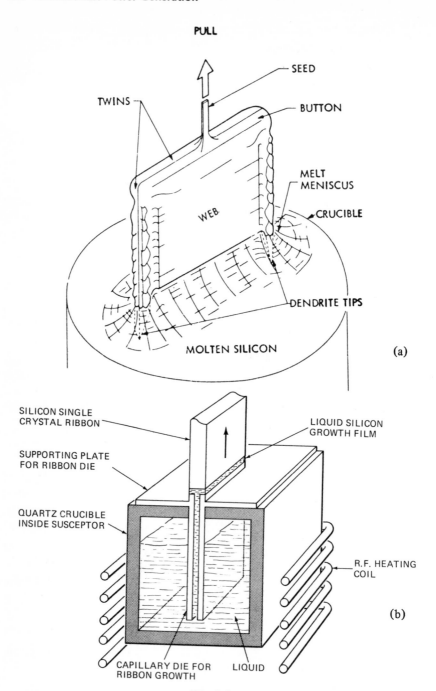

Fig. 4.6

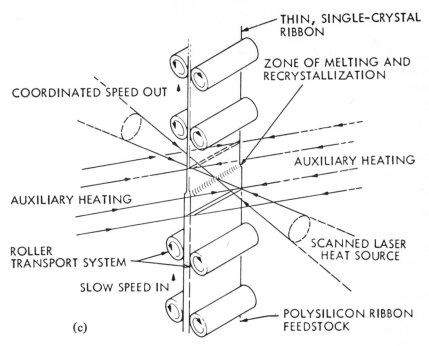

Fig. 4.6. Schematic drawings of (a) web-dendritic ribbon growth. (After Zoutendyk[201] with permission); (b) edge-defined, film-fed ribbon growth. (After Bates et al.;[204] courtesy of IEEE); (c) ribbon to ribbon growth by laser zone recrystallization. (After Zoutendyk[201] with permission)

Of the various crystal growth methods that use shaping dies the edge-defined film-fed growth method (EFG) is by far the most developed.[203] In this method, which is illustrated in Fig. 4.6b, capillary action serves to feed the molten silicon up the die slot, and in order to obtain reasonable pull-rates wetted-dies must be used. In the closely-related Stepanov technique, which uses nonwetted dies and hence has the potential of producing less contaminated crystals, an inverted geometry may prove necessary to utilize hydrostatic forces to obtain reasonable growth rates.[201] There is little information available on this technique yet, in contrast to the EFG process which has been under development for a number of years. EFG ribbons 2.5 cm wide and 250–300 μm thick have been grown in continuous lengths up to 24 m, but only small segments of the ribbon have been found capable of producing medium efficiency solar cells, e.g., 1 cm X 2 cm pieces have yielded 12% (AM1) cells, but this figure drops to 7% when utilizing 2.5 cm X 10 cm sections.[203] Even when used as substrates for epitaxial $N^+/N/P/P^+$ structures, 0.75 cm^2 areas of ribbon have so far yielded only 10% efficiencies.[205] If the EFG method is to be a commercial proposition then, besides improving the quality of the material, estimates show[305] that it will be necessary to increase ribbon width (to ~5cm), speed up the growth rate (to ~450 cm/hr), decrease ribbon thickness (to ~150 μm) and arrange for the growth to proceed automatically and feature continuous melt replenishment. One of the major problems in EFG ribbon growth is incorporation of transition metal impurities from the die and in order to reduce such impurities current technology subjects graphite dies to a high temperature purification process in HC1 ambients. The graphite crucibles also promote SiC inclusions which have been proven to be detrimental to the electrical characteristics of subsequently-formed solar cells.[206] Other defects occuring in EFG ribbons include linear parallel boundaries and intersecting boundaries, with the former dominating and being composed of twins of orientation (110) {211}. This structure also occurs in material formed by ribbon-to-ribbon growth (Fig. 4.6c) and suggests that for both growth methods the equilibrium structural characteristics could be determined by deformation-induced damage due to stress relief in the exit zone on solidification.[207] The incorporation of auxiliary heaters in the immediate postgrowth region might alleviate this problem. In the ribbon-to-ribbon technique the polycrystalline feedstock ribbon can be obtained from cast polysilicon material and thus some of the impurity problems associated with EFG ribbon can be avoided. The process is apparently capable of growing 127 μm thick single crystal ribbon at 12 cm min^{-1} [208] but the laser zone-melting technique shown in Fig 4.6c is probably very energy-intensive, and also the costs and times involved in the feedstock ribbon formation are probably far from negligible.

As was indicated in Section 4.1.4, it may not in fact be necessary to go to the trouble of preparing single crystal silicon in order for the required cost and efficiency goals to be met. Practical evidence to support this statement has recently been presented in the shape of granular silicon solar cells with AM1 efficiencies in the range of

10–14%.[187,188]   For these cells 350–450 μm thick slices were cut from precast blocks of Si composed of essentially fibrous grains with grain sizes of up to several millimeters. The term granular, or "semi-crystalline", has been used to describe this silicon material in order to draw attention to its properties of high order within the grains and of defects which can be assigned almost completely to the grain boundaries.[188]   These properties, as well as the large grain size, distinguish this material from the usual polycrystalline silicon.

Preparation methods for this new material are still proprietary, but presumably involve the remelting of ordinary polycrystalline silicon, followed by recrystallization in a suitable mould under favorable gas and temperature conditions.   The method has many attractive aspects, not least of which is the ability to form large area shapes amenable to high density packing and in a manner that is compatible with high volume production and automated fabrication techniques. Already modules comprising eight 10 cm × 10 cm solar cell elements have been constructed and found to be capable of delivering 6.5 W under AM1 conditions.[34]   There are indications that granular-Si cells are stable up to 500°C, that preferred orientations can be produced, that surface texturing is possible, that grain boundaries do not introduce undesirable shunting effects and that the material could be less sensitive than single crystal Si to secondary impurity inclusions.

All these factors suggest that a new and very important form of Si has been developed, and that modules made from this material could go a long way towards meeting acceptable cost, efficiency and production goals.   Further attention must be paid to reducing the current costs associated with slicing, contact technology and, perhaps, junction formation.   It is particularly encouraging that this material may be compatible with the use of reduced-grade silicon. Apparently there is evidence of grain boundary segregation of impurities, suggesting that the presence of a number of grain boundaries is in fact desirable, and that there may be an optimum finite size of grain.[188]

4.2.4 Silicon film preparation

Because of its absorption and minority carrier properties, silicon is not of much use as a solar cell material unless its thickness is in excess of about 50 μm and its grain size, if the material is polycrystalline, is greater than about 100 μm. In deposited-film technologies the substrate plays an important role in determining grain size, and for solar cell films a metallic substrate, which acted not only as a support but also as either an ohmic contact for a frontwall cell or a rectifying contact for a backwall cell, would be desirable. Many metals have been considered for this purpose[209] but none has proved able to foster silicon growth of the required grain size and structure.[210,211] Substrates with properties more akin to those of silicon are clearly required and it is this realization that has led to investigation of the chemical vapor deposition of silicon onto MG-Si and carbon substrates.[212,213]   100 μm crystallites have been formed in 200 μm thick plates of recrystallized P^+–MG-Si, and epitaxial p-n junctions (total thickness about 30 μm) have been deposited by the thermal reduction of trichlorosilane to yield 6.2% AM1 efficiencies for 30 cm^2 areas.[213]   Moderate fill factors (0.71) and open circuit voltages (530 mV) have been attained but short-circuit photocurrents are low (15 mAcm^{-2}). This is perhaps to be expected on account of the thinness of the active region. Furthermore the process technology currently used is not of a low cost nature owing to the need for SiHCl$_3$ and expensive contact metallization.

If the use of SiHCl$_3$ is considered acceptable to large-area, high-volume production techniques then it may prove economical to employ the proven epitaxial process, that has yielded 40 μm thick 12.6% efficient $N^+/N/P/P^+$ cells,[151] in conjunction with a reusable, high quality SeG-Si single crystal substrate. This might possibly be achieved by utilizing the peeled-film technology approach as illustrated in Fig. 4.7a.[214,308] The intermediate Si$_x$Ge$_{1-x}$ alloy film shown in the figure has a lower melting point than that of the Si alone and so the epitaxially-grown solar cell could, in principle, be peeled from the molten layer, thus allowing multiple use of the expensive seed

---

Fig. 4.7. Growth of Si films. (a) Peeled film approach using intermediate lower melting-point layer; (i) with riders resting on seed for Si cell growth; (ii) with riders raised and thin film cell peeled. The riders also move slightly inward, to allow bowing of the film during peeling (not shown). (After Milnes et al.[308] with permission). (b) Dip-coating approach, SiC coating ensures growth on one side only of ceramic substrate. (After Zoutendyk[201] with permission)

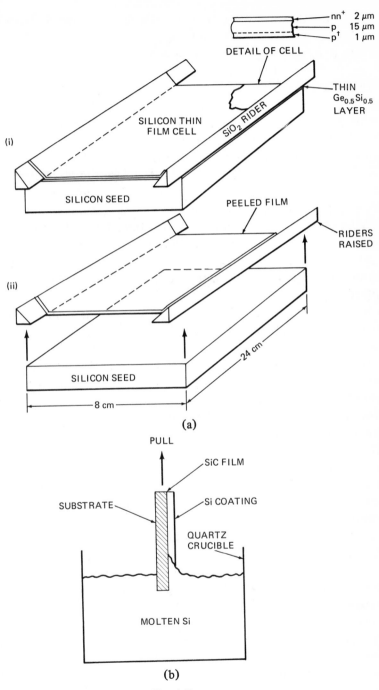

Fig. 4.7

block. Although no results have yet been reported for Si using this technique, 20 μm thick GaAs and $Ga_{1-x}Al_x$As-GaAs cells have been constructed in this manner, yielding AM2 conversion efficiencies of 11% and 13% respectively.[215]

Another fabrication technique involving silicon deposition via trichlorosilane utilizes Ti-coated alumina substrates and a covering layer of borosilicate glass and silicon dioxide.[216] On heating this sandwich above the silcon melting point dendritic silicon growth can be encouraged by unidirectional recrystallization. The Ti coating improves the Si adherence to the substrate (although its use with molten Si is not recommended, Section 4.2.1.), and the top glass layers prevent "balling-up" of the molten Si. Epitaxial deposition of P/N junctions on the dendritic silicon films has yielded solar cells of 3.2% efficiency.[216] The V_{oc} values were very low (0.32 V) presumably due to the unfavorable grain structure. Rheotaxial methods could presumably be employed to improve the Si substrate film grain structure,[217] but the number of deposition steps required and the dearth of suitable materials compatible with molten silicon makes these latter two methods of doubtful use for future solar cell fabrication needs.

A more promising approach to silicon films for solar cells is the dip-coating process illustrated in Fig. 4.7b. Primary crystallization of the silicon film actually occurs at the solid silicon-molten silicon meniscus interface, and so the grain structure is not primarily determined by the substrate. This has allowed the use of cheap ceramic substrates (carbon-coated to improve "wettability"), and 100 μm thick layers with 2–2.5 mm longitudinal grains have been attained with pull rates of 6 cm min^{-1}.[218] Present diffused junction cells are 4% efficient and are perhaps affected by impurities resulting from dissolution of the ceramic substrate.

The silicon referred to so far in this book has been either mono-crystalline or poly-crystalline, but silicon can also be prepared in an amorphous form in which the customary tetrahedral bonding is preserved for nearest neighbors, but a long-range disorder develops which is in contrast to the periodic lattice of crystalline silicon. Amorphous silicon (a-Si) can be prepared by either electron-beam evaporation, r.f. sputtering or deposition from silane in an r.f. discharge.[219] The optical properties of the resulting films are markedly dependent on the method of preparation with sputtered and evaporated films showing strong absorption out to wavelengths as high as 2 μm, and thus being of considerable interest for photothermal applications. The gas-discharge-grown films show promise for photovoltaic usage as in this case the optical bandgap is around 1.55 eV and the absorption coefficient is about an order of magnitude higher than that of crystalline silicon over much of the visible range. Cells as thin as 1 μm thus seem capable of AM1 conversion efficiencies around 15%.[220] In gas-discharge-grown films the low density of structural-defect localized states permits substantial doping,[219] and ~1 μm thick p-n and p-i-n junctions have already been fabricated.[220,221] The latter were designed specifically as solar cells and yielded 2.4% AM1 efficiency. Increases in this value to 5.5% have resulted from using Schottky barriers on a-Si,[222] and, if the hole transport properties could be improved,[306] the great potential of this relatively new material could perhaps be realized. The preparation procedure is simple and amenable to large-area coverage, substrates can be of stainless steel or coated glass, the films are truly thin and the amount of silane required is small.

4.2.5 Device processing techniques

With the advent of the cheapest possible SoG-Si and suitable means of preparing it into wafer, sheet or film form, further cost reductions in device fabrication must originate in the subsequent diode and solar cell processing steps. Methods of forming junctions, contacts and antireflection coatings must be sought which will replace some current procedures that are either costly, time-consuming or energy-intensive, e.g., vacuum deposition, photolithography, high temperature oxidation and diffusion.

The application of screen printing techniques to back and top contact metal deposition is a desirable move and currently receiving much attention.[3,173,174,223,304] Spin-on techniques are suited to antireflection film deposition, and a particularly interesting recent process involves the spinning of tetrabutyl titanate and the subsequent firing-through of the screen-printed top contact conductor pattern.[174] Instead of using antireflection coatings a textured silicon surface might be used. Such surfaces can be obtained using common and easily-manageable chemicals,[121,223] but the care and precautions that need to be taken in the successful processing and handling of such devices,[224] may preclude their use in nonconcentrating arrays.

Present commercial Si p-n junction cells use conventional high temperature deposition and diffusion techniques for the junction formation; some alternatives to the regular process are lower temperature diffusion from spun-on or doped-oxide sources, and ion-implantation. The former two methods allow simultaneous formation of both back and front surface layers and look promising from an automation standpoint.   Using spun-on or CVD-deposited doped oxides 14% (AMO) efficient cells have already been realized,[225] whilst with non-oxide spun-on sources, there is evidence that the expected problems of nonuniform deposition, bubble formation and cracks on drying, can be overcome in large-area junction fabrication.[226]   Complete device fabrication at temperatures lower than about 400°C is possible with ion-implantation techniques using either corona[227] or charged-beam[88] ion sources.   In one development of the latter method pulsed electron beam techniques have been perfected to allow room-temperature annealing of the implant damage,[88] and 2 cm × 2 cm 10% (AMO) efficient cells have been fabricated in less than 2 minutes! With the advent of high-current ($\sim$100 mA) ion implanters a production facility can be envisioned (Fig. 4.8) capable of supplying 2×10^8 peak output watts per year of solar cells at an estimated cost, excluding the starting material, of about \$0.20/W.[88]

Low temperature device processing is desirable in order to minimize degradation of minority carrier properties but, in addition to meeting this requirement, it would be advantageous if junction formation methods could avoid the creation of surface "dead layers," which can result from diffusion- or implant-induced crystal damage. Yet another desirable attribute of the junction formation procedure would be its avoidance of drive-in dopant procedures that may cause preferential-diffusion down grain boundaries, and thus render the method unsuitable for use with some polycrystalline materials. All these criteria can be met by utilizing surface junction devices, examples of which are Schottky barriers (MS and MIS), grating cells, glassy metal oxide cells and inversion layer (IL) devices (see Fig. 4.9). The performance of some of these cells was discussed in Chapter 3. One

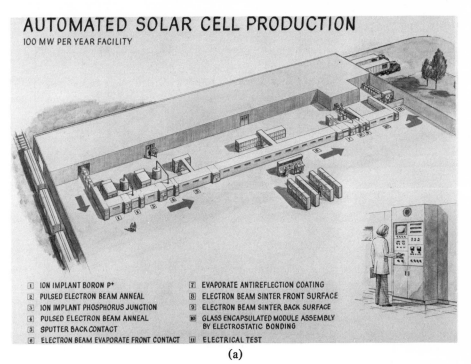

AUTOMATED SOLAR CELL PRODUCTION
100 MW PER YEAR FACILITY

1 ION IMPLANT BORON P⁺
2 PULSED ELECTRON BEAM ANNEAL
3 ION IMPLANT PHOSPHORUS JUNCTION
4 PULSED ELECTRON BEAM ANNEAL
5 SPUTTER BACK CONTACT
6 ELECTRON BEAM EVAPORATE FRONT CONTACT

7 EVAPORATE ANTIREFLECTION COATING
8 ELECTRON BEAM SINTER FRONT SURFACE
9 ELECTRON BEAM SINTER BACK SURFACE
10 GLASS ENCAPSULATED MODULE ASSEMBLY BY ELECTROSTATIC BONDING
11 ELECTRICAL TEST

(a)

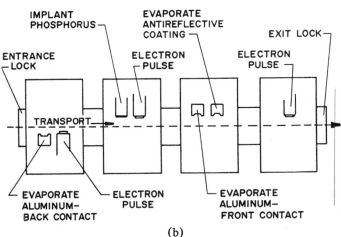

(b)

Fig. 4.8. (a) High-volume production facility concept for Si N/P cell fabrication. (Courtesy of A. R. Kirkpatrick, Spire Corp.) (b) Detail of ion-implantation and pulsed-electron beam-annealing stations (After Kirkpatrick;[88] courtesy of Amer. SES).

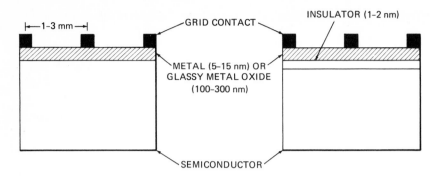

SCHOTTKY BARRIER OR GLASSY METAL OXIDE HETEROJUNCTION CELLS

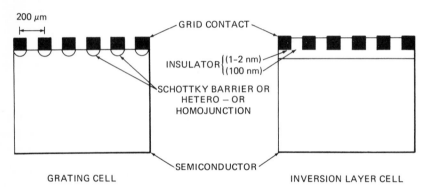

Fig. 4.9. Surface junction solar cells.

not mentioned specifically was the IL cell, which is similar in operation to that of the earlier grating cell,[228] and utilizes the charge trapped in a thin covering oxide to induce an inversion layer over the whole semiconductor surface.[229-31] By combining this effect with metallization and insulator thickness appropriate to minority carrier MIS diodes, then some advantage over the latter devices might accrue from reduced photon reflection. In these MIS and IL devices induced junction formation is facilitated by using high resistivity base material but, as with p-n junctions, this leads to a potentially low V_{oc}. However, because the electrical properties of these diodes are akin to those of p-n junctions, back surface fields can be employed to boost V_{oc}.

Possible drawbacks to thin insulator devices concern the obtaining of uniform oxides over large areas and the deterioration in device properties due to instabilities in the oxide.[232] Degradation of some $SnO_2/SiO_x/Si$ cells has been attributed to changes of this latter nature.[233] Indium-tin-oxide/Si devices may be less susceptible to this phenomenon on account of the more favorable relationship that exists between the chemical free energies of formation for the metal oxide and silicon dioxide.[124]

In all the above devices some understanding of the factors affecting solar cell operation has been gathered by utilizing single crystal substrates, and conversion efficiencies around 10% have been attained. The next step in their development would seem to be to assess performance using nonsingle-crystalline substrates. Some progress in this direction has been reported with $SnO_2/polySi$ devices, and 6.9% efficiencies have already been achieved on 4 cm^2 devices.[234]

4.3 CADMIUM SULFIDE SOLAR CELLS

Until recently the term "cadmium sulfide solar cell" invariably implied a Cu_2S/CdS heterojunction with the CdS layer acting as the solar cell base. Photovoltaic action was first observed with this structure in 1954,[235] and considerable development since that time has culminated in a stable, reasonably reproducible device with AM1 conversion efficiency approaching 10%.[113] Whilst this efficiency level and the relatively simple cell fabrication procedures will almost certainly lead to the use of Cu_2S/CdS cells in terrestrial applications, it is possible that their use in the large system scenarios discussed in Chapter 2 will be precluded by prohibitive real estate and electrical interconnection costs.

The limitation to Cu_2S/CdS cells as regards efficiency is the nature of the semiconductor interface, not the CdS material which forms an attractive window material in that it can be inexpensively deposited in suitable thin film form on a number of convenient substrates. To improve interfacial properties the surface layer of Cu_2S can be retained, but used in conjunction with Zn-doped CdS, or alternatively,

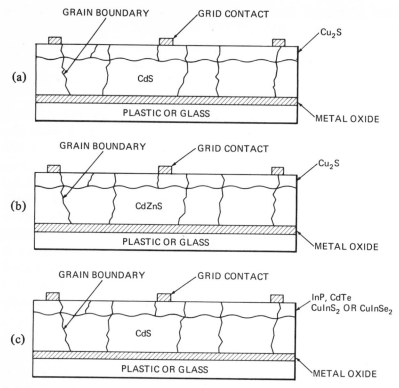

Fig. 4.10. Thin film heterojunction solar cells based on CdS. (Adapted from Dubow[3] with permission).

different absorber materials can be substituted for Cu_2S (Fig. 4.10). CdTe, InP, $CuInSe_2$ and $CuInS_2$ are potentially suitable materials that could lead to the emergence of a new "cadmium sulfide cell."

4.3.1 Cu_2S/CdS cells

Until very recently the only commercial method of fabricating Cu_2S/CdS solar cells was based on that developed by the Clevite Corporation in the United States and the type of cell resulting from this process is illustrated in Fig. 4.10a. The fabrication sequence consists of the vacuum evaporation of CdS onto heated substrates of either metal or metallized plastic, followed by dipping into aqueous cuprous chloride solution at around $90°C$ thus forming a surface layer of Cu_2S by topotactic ion exchange. This is followed by attachment of a gold or copper grid contact system and finally a plastic cover. The resulting thicknesses of the Cu_2S and CdS layers are around 0.2 and 25 μm respectively and the device is often subject to various post-fabrication heat treatments to improve photovoltaic performance.[172] Although the dipping process is simple, inexpensive and suited to high volume production, difficulties have been experienced with this method in the form of solar cell instabilities and lack of copper sulfide stoichiometry (the chalcocite phase, Cu_xS with $1.995 \leqslant x \leqslant 2.0$, is required for good photovoltaic action).[236,237] Treatments have been devised to effectively eliminate the various stability problems[134,172,238,239] but the reproducibility of cells formed by the Clevite process is difficult to ensure owing to problems in controlling the dipping reaction. This has led to the consideration of alternative methods of Cu_2S deposition, namely: a dry chemical process involving the evaporation of cuprous chloride followed by heat treating at $180°C$ for about 2 minutes;[236] reactive sputtering;[240] sulfurization of copper in H_2S[241] or thiourea;[242] pyrolytic chemical spraying.[184]

Pyrolytic spraying is also suited to CdS deposition and dense, transparent films can be prepared on SnO_x-coated glass substrates. The CdS film properties depend markedly on substrate temperature and subsequent hydrogen annealing, but under favorable conditions films with electrical properties similar to those of evaporated films can be obtained.[243] In addition, the superior mechanical and structural properties of the spray-deposited films allow the use of films that are only 2–3 μm thick.[184] Such thicknesses would certainly reduce somewhat any doubts about the ability of Cd reserves to meet the demands that might result from widespread use of CdS-based solar cells. The spray deposition method is well-suited to high volume production techniques and one possible process is illustrated in Fig. 4.11. A plant based on this design has been described which would be capable of producing 37×10^6 m^2 of glass annually, in the form of 3 mm thick, 3 m wide ribbon.[184] Cells, with 5% conversion efficiency, produced on such substrates could apparently cost as little as $0.06/W.

One problem in utilizing arrays made with such low efficiency cells

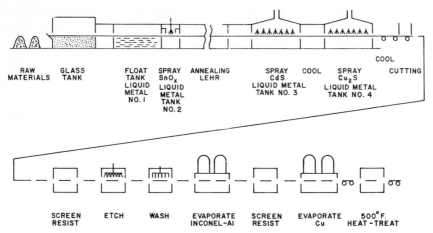

Fig. 4.11. Schematic drawing of augmented float glass plant for the automatic, high-volume production of Cu_2S/CdS solar cells. (After Jordan;[184] courtesy of IEEE)

is the number of interconnects required. A scheme to overcome this and maintain a high volume, automatable, production capability has recently been proposed.[244] Chemical spraying is again used for the semiconductor depositions and an integrated array is constructed utilizing indium shorting bars (Fig. 4.12). Apparently, evaporated indium layers reduce the underlying Cu_2S and the result is a low resistance contact between adjacent cells. Clearly this arrangement would give a very high packing density within an array, but the high mobility of the deposited indium poses some questions concerning the likely stability of the system.

The above processes based on spray deposition of Cu_2S and CdS are attractive because of their high volume capability, but the current cell efficiencies of around 5% are too low for all but the very small-sized (electrically) terrestrial applications. However, the expected low cost of Cu_2S/CdS arrays should ensure some penetration into markets represented by these applications. A plant capable of annually producing 500kW of Cu_2S/CdS cells by a modified Clevite process is being established in anticipation of supplying this market.[245] To obtain markedly higher conversion efficiencies with Cu_2S/CdS cells will probably require Zn doping of the CdS; such cells are still only at the experimental stage (Chapter 3).

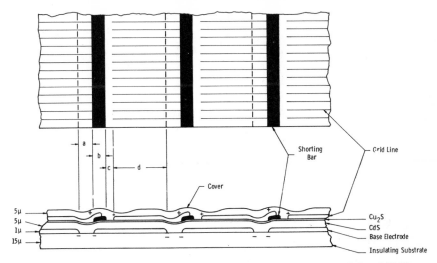

Fig. 4.12. Integrated array structure showing series connection of Cu_2S/CdS solar cells via In shorting bars. Not to scale. (After Shirland et al.;[244] courtesy of IEEE)

4.3.2 Other heterojunction cells using CdS

In seeking companion semiconductors to CdS for solar cell use the following characteristics would be regarded as desirable: direct bandgap with value around 1.0–1.6 eV; ability to be prepared in a low resistivity, thin film form; formation of a junction with CdS that had good lattice match and no band-edge spikes; inherent stability. Materials that would appear to meet all or most of these requirements are CdTe, InP, and various I–III–VI$_2$ ternary compound semiconductors. Heterojunctions based on single-crystal substrates of CdTe, InP and CuInSe$_2$, and using CdS window layers, have yielded AM2 efficiencies of 10.5%,[131] 15%[303] and 12%[132] respectively. These efficiency values demonstrate the potential of these materials, but in order to be cost-competitive in unconcentrated-sunlight applications it is desirable that their capability for thin film fabrication be realized.

Perhaps the most spectacular development in this direction is the recently-announced ceramic CdTe cell (Fig. 4.13) for which an AM0 efficiency of 8.1% has been reported.[246] This corresponds to an AM2 efficiency close to 10% and is remarkable in that it has been achieved

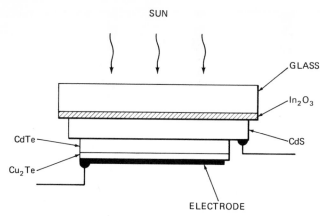

Fig. 4.13. Ceramic thin film CdTe solar cell. (Adapted from Nakayama et al.[246])

in a cell formed from semiconductor pastes and screen printing techniques. CdS and CdTe resistivities were reported as 0.2 and 0.1–1 Ω cm respectively and CdTe grain sizes were 2–10 μm. A key feature of this cell is the treatment used to form the p^+-Cu_2Te back contact layer. Not only does this layer allow simple In-Ga alloy electrodes to be applied but it also leads to some p-type doping of the CdTe layer such that the cell structure is probably n-CdS/n-CdTe/p-CdTe/p^+-Cu_2Te.[246] It therefore might be classed as a back surface field, homojunction CdTe cell with a heteroface structure. The CdS layer acts as a low resistance transparent ohmic contact and also most likely reduces the recombination velocity at the CdTe surface, in the same way that a $Ga_{(1-x)}Al_xAs$ layer does when deposited on top of GaAs junctions (Chapter 3). The Cu_2Te film was prepared by a similar dipping technique to that used for formation of Cu_2S on CdS. Whether this will lead to some of the stability problems that plagued early Cu_2S/CdS cells remains to be seen, but certainly the truly low cost nature of the cell processing and the demonstrated good efficiency merits further development of the ceramic CdS/CdTe cell.

Thin film cells of CdS/InP,[156] CdS/CuInSe$_2$[247] and CdS/CuInS$_2$[247] are also being investigated and present efficiencies are 5.7% (AM2), 6.2% (AM1) and 2.9% (AM1) respectively. Both p and n-type CuInS$_2$ can be readily prepared suggesting that homojunction cells in this material might be useful. Present AM1 efficiencies for such cells are 3.6%,[247] and as the reported short-wavelength spectral response is poor, it is possible that significant improvement in performance could result from incorporation of a CdS heteroface junction, as used in the CdTe cells described above. The stability of $CuInS_2$ seems good but there may be some doubts about $CdS/CuInSe_2$ cells in this regard as Cd migration into the $CuInSe_2$ has been observed at 200–300°C.[247] This may present a limitation during device processing and, should migration also occur at lower temperatures, would also limit the suitable operating environments. Efficiencies in present $CdS/CuInSe_2$ cells seem to be limited by low values of V_{oc} and FF, both of which may be associated with poor interfacial properties or small grain sizes. Similar problem areas may pertain to thin film CdS/InP cells, for which current values of V_{oc} and FF are about 40% and 15% lower, respectively, than have been obtained in single crystal devices.[156] An interesting feature here is the spectral response of the thin-film cells which indicates an absorption tail on the InP side, the greater extent of which seems to be associated with larger values of V_{oc}.[156] Further understanding of the polycrystalline InP material is required, as is a suitable method for heavily doping this material and so obviating the need to use intermediate low resistance contact layers between the InP and present graphite substrates.

4.4 OTHER SEMICONDUCTOR POSSIBILITIES

The factors that have led to the dominant positions held by Si and CdS as basic semiconductor materials for large-area photovoltaic conversion systems have been presented in this chapter. The sheer extent of the developments and investments that have been made in the technologies of these materials assures them of exposure to the terrestrial solar cell market place. However, as the understanding of factors affecting solar cell performance develops, and the realization of the merits of photovoltaic power systems grows, other materials and combinations of materials may emerge as being suitable for photovoltaic power generation in unconcentrated sunlight. Some of the possibilities currently being explored are discussed below.

4.4.1 Inorganic semiconductors

Some semiconductor systems with potential suitability to large area photovoltaics and that have not been discussed thus far in this chapter are: CdSe/ZnTe, ZnSe/CdTe, and ITO/CdTe heterojunctions;[157] GaAs heterojunctions[128] and Schottky barriers;[155] Cu_2O Schottky barriers;[248,249] devices based on Zn_3P_2;[250] $ZnSiAs_2$;[250] $CuGaSe_2$;[251] $AgGaSe_2$;[251] $GaAlSb$[252] and other II–VI–V_2 and I–III–VI_2 compound semiconductors.[252a] No results have yet been reported for solar cells based on the last group of materials.

Of the II–VI heterojunctions listed above those involving CdTe are of particular interest as this material has a demonstrated thin film solar cell capability.[246,253] However, practical conversion efficiencies in ZnSe/CdTe and glassy metal oxide/CdTe cells are currently very low and indicate the presence of conduction band spikes or severe interfacial recombination.[157] Some of the practical problems associated with CdTe have been identified with extremely short minority carrier lifetimes (10^{-8}–10^{-9} sec), presumably due to trapping levels within the bandgap related to structural defects.[254] Trap reduction using complexing dopants, and the use of narrow bandgap, defect conducting, transition metal tellurides in heterojunction structures, have been proposed as measures that might lead to improved CdTe devices.[254]

Thin film Schottky barrier GaAs cells were reported in 1967 with AM1 efficiencies of 4.5% for 2 cm² areas.[91] To improve efficiencies to around 12% would, in theory, require an MIS configuration and at least 10 μm grains with a columnar structure.[102] Recent evidence suggests that the structural requirements can be met by the close-spaced vapor transport of GaAs onto Mo substrates.[255] The cross section of a potentially suitable film is shown in Fig. 4.14, but the use of uncoated Mo substrates is not suited to solar cell systems on account of the mildly-rectifying nature of the GaAs/Mo contact.[91, 255] Evaporation of Ge onto metal foils and subsequent recrystallization of the deposited film is a method that shows promise of providing both a low resistance contact to GaAs and a suitable seed plane for millimeter-sized grains.[155] Pending the solving of the substrate problem thin film MIS GaAs cells have been prepared on polycrystalline GaAs substrates, and the present indications are that device perform-

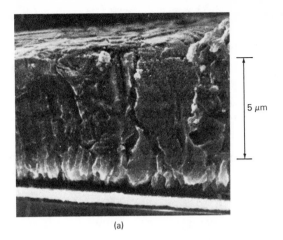

(a)

(b)

Fig. 4.14. Cross-sectional SEM photomicrographs of solar cell thin films. (a) GaAs deposited on Mo by close-spaced vapour transport. (After Russel et al.;[255] courtesy of IEEE); (b) thermally-oxidized Cu_2O. (Courtesy of V. F. Drobny[259]). Grain boundary-revealing etches used: (a) $HNO_3:HF:H_2O$; 1:1:1; (b) $HNO_3:H_2O_2:H_2O$; 0.25: 0.09:0.66.

ance is limited only by the nature of the interfacial oxide layer, and changes in barrier height that occur on deposition of antireflection coatings.[155] Work being carried out on single crystal GaAs MIS cells may help characterize the interfacial oxide properties, which at present seem to depend on the method of oxide formation, and the preparation and orientation of the GaAs surface.[155] The presence in polycrystalline films of variously oriented surfaces, some of them unfavorable as regards desirable oxide formation, may be the reason for the reduced photovoltage response of current devices.[155] Another promising approach to thin film GaAs cells utilizes the AlAs/GaAs heterojunction which, in its single crystal form, has attained an AM1 efficiency of 18.5%.[106] A thin film version of this cell deposited on high density graphite using chloride transport vapor phase epitaxy appears to have good interface, ohmic contact and stability properties.[128] Current problems involve reproducibility (associated with substrate preparation) and series resistance (due to poor lateral conduction between grains in the top AlAs layer). As thicknesses not much in excess of 2 μm are satisfactory for GaAs cells there would not appear to be any cause for concern over the ability of Ga and As reserves to meet the demands imposed by a successful GaAs cell technology. In fact the crustal abundance of these two elements has been estimated to exceed that of Cd, for example, by 100 and 10 times respectively.[256]

One of the earliest semiconductors demonstrating photovoltaic action was Cu_2O, and interest in this material has recently been revived, principally on account of the very simple thermal oxidation operation required to convert Cu to Cu_2O.[248,249] Partial oxidation of the Cu forms a ready-made Schottky barrier at the Cu_2O/mother Cu interface but the barrier height is only 0.7 eV ($\sim E_g/3$),[257] so an MIS structure is needed to obtain reasonable conversion efficiencies.[248] The growth and annealing times and temperatures required for large grain Cu_2O formation depend on the original Cu sheet thickness;[258] foils above about 500 μm thick would not appear compatible with low cost solar cell production as annealing times approaching 100 h are necessary. For thinner foils the high temperature processing times are more reasonable and Fig. 4.14b shows centimeter-sized grains of columnar structure obtained on subjecting a 200 μm Cu foil to 5 hr in air at 1040°C.[259] As-grown film resistivities are high, around 5000–30,000 Ω cm, and this may pose a problem as doping of Cu_2O is apparently not easily accomplished.[260]

Progress needs to be made in this area and in ohmic contact and junction formation before any firm evaluation of the photovoltaic importance of this material can be made. With a bandgap of 2 eV the photovoltage response can be expected to be high and the photocurrent response low. It is possible that Cu_2O could find a place in any scheme that sought to arrange solar cells made from different materials in a tandem arrangement, in order to utilize a larger part of the solar spectrum than a single cell is able to do. Two possible schemes are illustrated in Fig. 4.15. Although theoretical conversion efficiencies can reach 35% for suitable combinations of cells,[261, 262] the practical problem of electrical matching, particularly when differ-

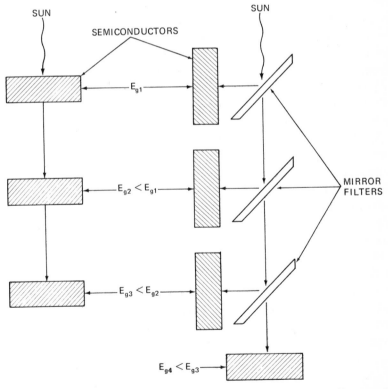

Fig. 4.15. Possible arrangements for operation of dissimilar solar cells in tandem.

ent cells will be affected differently by changes in insolation, would appear to be a serious drawback to the scheme.

4.4.2   Organic semiconductors

Organic materials can be formed very cheaply in large-area, thin film form; some organic films are semiconducting and diode action can often be achieved via a Schottky barrier. These facts are sufficient to warrant some study of the possibilities of organic solar cells. Materials that have received some attention are:[263-6] anthracene, tetracene, chlorophyll, various phthalocyanides, a 1:1 complex of poly (N-vinylcarbazole) and trinitrofluorenone (PVK-TNF), and squarylium dye. All these materials are highly resistive (10^5–10^8 Ω cm) and exhibit a somewhat variable and not always instantaneous photovoltaic effect by virtue of a Schottky barrier being formed at one, or both, of the contacted surfaces. Because of the high resistivity, films must necessarily be thin, but as absorption coefficients are generally high in the materials mentioned above, this does not constitute a serious photon absorption problem. For the PVK-TNK system in films greater than 1 μm thick the conversion efficiency is apparently limited to less than 0.1% by the onset of space-charge limited conduction. At thicknesses less than this the theoretical conversion efficiency can reach 1%, provided the photon absorption can be kept high by introducing, for instance, highly absorbing pigments.[265] Theoretical treatments are only tentative at present though, on account of the uncertainty of the band structure in many of these materials, the presence of large trap densities and the precise influence of the metal-semiconductor contacts.[264,265,267]

The highest practical efficiency (AM0) so far recorded is 0.1% at a solar intensity of 5 mW cm^{-2} (which drops to 0.02% at 135 mW cm^{-2} on account of series resistance effects),[266] although there are rumors of a 1% efficient organic cell.[171] The series resistance problem stems from the use of such high resistivity material and the difficulty this presents in making ohmic contacts.

Future work, besides seeking new materials, is likely to concentrate on improving the quantum yield by reducing trap densities in the films; this could be achieved by purification and solidification procedures which might then pave the way for intentional doping so allowing a reduction in series resistance and improvement in ohmic contact properties.

4.5 ENERGY PAYBACK TIME

It is possible that several of the solar cell materials or material combinations discussed in this chapter may come close to meeting desired cost goals in terms of the $/W index. As the end product from the solar cell is energy, it is also important therefore, that solar cell production methods not be overly energy intensive. Of the various methods discussed in this chapter, details on energetics are only available for some instances of silicon device processing.   For this material the ion-implantation scheme shown in Fig. 4.8 is very conservative energetically, and it has been estimated that all the energy used to process 10% efficient cells could be "repaid" in the course of one operational day.[88]   This can be compared to energy payback times of 6 months that have actually been measured for conventional modules (not just cells) manufactured by Solarex Corp., United States.[268] These payback times cover energy usage during the conversion of a silicon slice to a solar cell (or module) and do not include energy expended in producing the silicon in the required form. The payback time of this latter energy has been estimated as 3 years,[268] although some recent calculations are more pessimistic and put the total energy payback time at 12 years.[191] However by avoiding the energy-intensive $SiHCl_3$ process for producing SeG-Si and reducing kerf losses, the *total* energy payback time could be as little as 4 months;[191] this seems perfectly acceptable for a product with an expected lifetime of 20 years. These energy payback times will, of course, be even shorter if operation is under conditions of concentrated sunlight, e.g., it has been predicted that some Si modules designed for use at concentration values approaching 1000 could return their energy investment in as little as 1–5 weeks.[301]

5.

Solar Cells for Concentrated Sunlight Systems

In the concentrated sunlight approach to photovoltaic energy conversion the high dollar cost of solar cells is taken as an accepted fact, but the $/W index is reduced by increasing the cell output power density. For a given solar cell conversion efficiency and output power the collector area required is, of course, the same whether unconcentrated or concentrated sunlight is used, but with the latter approach a large fraction of the collecting area comprises mirrors or lenses, which are, perhaps, more amenable than solar cells to low-cost, high-volume production. With the solar cell freed from the design constraints of low-cost and large-area compatibility, full attention can be given to realizing high values of conversion efficiency. Design trade-offs that lower cell conversion efficiency are not likely to be cost-effective in the concentrated sunlight approach because of the associated increases in land, concentrator area and support and tracking structure that would be required. This statement becomes increasingly more relevant as the sunlight concentration ratio is raised. For illumination levels less than about 10 Suns (1 W cm^{-2}) the solar cells described in Chapter 4 are probably suitable, but different designs are required if high values of conversion efficiency are to be maintained at higher values of concentration ratio and, possibly, elevated temperatures.

Conversion efficiency increases, up to a point, with increasing illumination intensity, but solar cell performance is degraded by temperature increases much in excess of usual ambient values. The principal factor affecting high temperature operation is the associated increase in intrinsic carrier concentration which leads to larger diode dark currents and reduced values of open circuit voltage and fill factor. The use of high bandgap materials mitigates this effect but cooling of terrestrially based cells is not a great problem so that materials with low bandgap, e.g., Si, are not automatically excluded. In some instances the withdrawn thermal energy may be usable in space heating and cooling or in low temperature industrial processes, so permitting a high degree of utilization of the incident solar energy. Some of the factors governing the operation of photovoltaic systems at high levels of concentration were discussed in Section 2.2.2. and, as far as solar cell performance is concerned, the problem of maintaining a high conversion efficiency is principally one of reducing the series resistance so that the power losses under high photocurrent conditions do not become excessive. Reducing R_s is mainly a technological problem involving junction depth and top contact metallization, but J_P is higher in lower bandgap materials and so, again, high bandgap materials appear favorable for operation in concentrated sunlight.

Of the currently successful solar cell absorber materials the one with the highest bandgap is GaAs, and so the interest in its use as a base material for concentrator solar cells is understandable. Cells with the structure p-Ga$_{(1-x)}$Al$_x$As/p-GaAs/n-GaAs/n^+-GaAs have already been constructed that yield conversion efficiencies of 23% and 19.1% at concentration ratios of 10 and 1735 respectively. [170] Further development of concentrator systems using this material is thus likely to be concerned with integration of the cells into arrays and with the problems of obtaining low-cost optical concentrators and sun-tracking apparatus.[51] The fabrication of solar cells from absorber materials with higher bandgaps than GaAs has been very limited (see Section 4.4.1. and Refs.[252a, 270]). The other solar cell materials that have realized high conversion efficiencies, or seem capable of doing so, all have lower bandgaps than GaAs and thus, for use in concentrated sunlight environments, they must be well-cooled and be compatible with technologies and designs that can keep R_s at an acceptable value. The only material that presently seems able to meet these requirements is silicon. Using p-n homojunction diodes conversion efficiencies of 19%,[122] 13.5%[269] and 12.8%[271] have been realized at illumination levels of 1, 50 and 109 Suns, respectively. The last two results were obtained using cooled cells with conventional diffused junctions and elaborate patterns of top contact metallization. Other approaches to solving the R_s problem in silicon have involved the use of geometrical arrangements other than the simple planar junction structure. The increased complexity of the latter designs is justifiable if operation at very high illumination levels (greater than several hundred Suns) is made possible.

The features of silicon and gallium arsenide cells that have been designed for use in concentrator systems are described in this chapter, following a brief discussion of some of the influences that operation in concentrated sunlight exerts on solar cell conversion efficiency.

5.1 CONVERSION EFFICIENCY IN CONCENTRATOR SOLAR CELLS

As the prime requirement of concentrator solar cells is a high conversion efficiency it is important that the factors affecting solar cell performance, as described in Chapter 3, be considered in any successful cell design. In addition the cell performance must be evaluated under conditions of high illumination intensity and, quite possibly, elevated temperatures also.

5.1.1 High illumination intensity effects

In a solar cell that exhibits a single exponential dark I–V characteristic (with saturation current $I_o \ll I_p$) and for which effects due to shunt and series resistance are negligible, it is easily shown that

$$I_{sc} \propto P_i \quad \text{and} \quad V_{oc} \propto \ln P_i \quad (5\text{-}1)$$

where P_i is the input solar power density. If the fill factor of the device is not affected by P_i then it follows from Equations 5.1 and 3.2 that the conversion efficiency can increase with illumination intensity by an amount, in the ideal case, directly proportional to the increase in photovoltage. For low levels of injection (photogenerated minority carrier density $\ll$ initial majority carrier density) it is generally accepted that the photovoltage saturates at a value very close to the junction built-in voltage or diffusion potential V_d,[149] and so unlimited increases in conversion efficiency cannot, of course, be expected. Also, as solar cells are usually asymmetrically doped with the base region being of greater resistivity than the surface layer, conditions of high level injection in the base are likely to arise before the condition $V_{oc} = V_d$ is reached. Under these circumstances charge neutrality in the base can only be maintained by an increase in majority carrier concentration to match the new value of minority carrier concentration. This implies that the quasi fermi levels in the base can no longer be regarded as flat and so, even on open circuit, there is a potential drop across this region leading to a reduction in photovoltage below $V_j = V_d$.[272] If current is allowed to flow in a diode subject to high injection conditions the dark I–V characteristic becomes roughly proportional to $\exp(qV_j/2kT)$ with a saturation dark current value greater than that which would apply for low level injection conditions.[138] This phenomenon would thus tend to lead to a reduction in fill factor.

In practical solar cells, on account of the presence of a finite series resistance, a reduction in conversion efficiency may well occur at illumination levels below that at which the above effects might be manifest. Series resistance is undesirable in all solar cells, but particularly so under conditions of sunlight concentration where the current levels are high. Figure 5.1 demonstrates this by presenting calculations of the fill factor for a silicon diode, as computed from Equation 3.26 for various values of concentration ratio C. It can be seen that for satisfactory operation at 100 Suns, e.g., R_s must be less than a few hundredths of an ohm. For GaAs cells somewhat higher values of R_s would be tolerable on account of the lower photocurrent response of this material. To calculate the data shown in Fig. 5.1 it was assumed that the photocurrent increased linearly with the input power density, although it has recently been suggested that saturation of I_P (at a value given by V_d/R_s) may occur at high illumination levels.[273] Such an effect would delay somewhat the fall-off in FF shown in Fig. 5.1, but the overall conversion efficiency would not be improved because of the reduced photocurrent. If conditions of high level injection were appropriate then the associated increase in majority carrier concentration would lead to a modulation of the semiconductor conductivity, in which case R_s would decrease as P_i increased, and so the

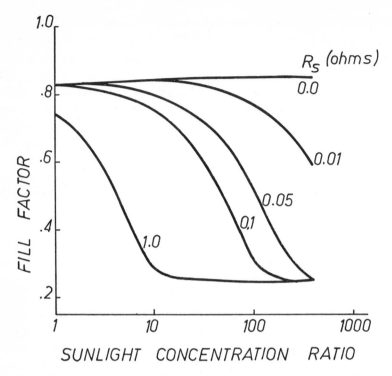

Fig. 5.1. Fill factor variation with sunlight concentration ratio for different values of series resistance. Calculated from eqn. 3.26 for Si assuming that $\gamma = 1.5$, $J_0 = 3.3 \times 10^{-12}$ A cm^{-2}, $a = 1$ cm^2, $T = 300°$K and $I_P \propto C$.

degradation in *FF* may be reduced somewhat. However, operation in the conductivity modulated mode still implies a reduced value of V_{oc} (through the dark current properties), and so, whilst it is possible that the conversion efficiency may fall less rapidly with P_i than is the case for cells operating under low level injection conditions, the initial values of η in the latter case are likely to be higher. To combine the desirable features of conductivity modulation with a high initial value of V_{oc} in a single device may be possible by utilizing a back surface field arrangement,[269] as can be seen from the results of detailed numerical calculations, as presented in Table 5.1. For GaAs cells the short minority carrier diffusion lengths preclude the use of back surface fields but, as doping densities are generally higher than in Si cells, high level injection effects are not likely to be important until higher values of sunlight concentration ratio are reached.

Table 5.1. Calculated performances of 10 Ωcm N^+/P and $N^+/P/P^+$ solar cells at 1 and 40 Suns AMO and 27°C and 100°C.[a]

T (°C)	X (SUNS)	BSF	J_{sc}(mA/cm^2)	V_{oc}(V)	FF	η(%)
27	1	NO	49.9	0.541	0.808	16.1
27	1	YES	52.0	0.606	0.809	18.8
27	40	NO	2130.0	0.605	0.715	17.0
27	40	YES	2130.0	0.721	0.755	21.4
100	1	NO	50.4	0.350	0.702	9.2
100	1	YES	52.6	0.438	0.717	12.2
100	40	NO	2110.0	0.432	0.625	10.5
100	40	YES	2110.0	0.580	0.714	16.1

[a]From Ref. 269.

5.1.2 High temperature effects

Although sunlight concentration can lead to increases of solar cell conversion efficiency, at least for low values of C, the maximum photovoltaic conversion efficiency for a single cell is still around 25%. This implies that there is considerable thermal energy at the collector which could lead to large increases in cell temperature if adequate heat-sinking is not provided. Such an effect would primarily affect the minority carrier properties and intrinsic carrier concentrations in the cell, and to a lesser extent the photon absorption properties. For the doping densities typical of Si and GaAs solar cells the minority carrier mobilities change only slightly (decreasing somewhat in the case of Si) with temperature, with the result that the minority carrier diffusion coefficient is nearly independent of temperature for Si, but increases monotonically with T for GaAs.[103] As minority carrier lifetimes usually increase with temperature, owing to increased thermal velocities, minority carrier diffusion lengths ($L = \sqrt{D\tau}$) increase with temperature, and particularly so in the case of GaAs. Larger diffusion lengths imply higher photocarrier collection efficiencies and thus an increase in J_{sc} with T. This increase in J_{sc} would also result in an improvement in V_{oc} with temperature were it not for the fact that V_{oc} is strongly dependent on the intrinsic carrier concentration n_i and this increases exponentially with T. The saturation dark current density is proportional to n_i^2 and n_i for the cases of injection-diffusion and recombination–generation currents respectively, so in diodes for which these current mechanisms are appropriate both the open circuit photovoltage and the fill factor will decrease with T. The slight decrease in bandgap energy with temperature also contributes to an increased saturation dark current and this may negate any beneficial effect that the associated improved long wavelength absorption might otherwise have.

Taken together, the above effects can be expected to produce a reduction in conversion efficiency once T is allowed to increase much above room temperature. For heterojunction cells in which the dark current is dominated by tunneling there could be less sensitivity to temperature increases than in the above cases, although problems of electrical stability would likely arise in current solar cell structures for which tunneling is an important conduction mode. For Schottky barrier solar cells an increase in T would be expected to be particularly undesirable as J_o is proportional to $T^2 \exp(-q\phi_b/kT)$ (Equation 3.25); however measurements on Si devices indicate a temperature dependence almost identical to homojunction cells.[274] The presence of an interfacial insulating layer in Schottky barrier or heterojunction solar cells is likely to degrade the performance of such cells in concentrated sunlight, owing to the associated series resistance.[275]

As the principal high temperature degradation mechanism in high efficiency solar cells appears to be the fall-off of V_{oc} with T, an expedient way of dealing with this problem would seem to be to make the initial value of V_{oc} as large as possible. The incorporation of a back surface field was shown in Chapter 3 to be effective in achieving this in high resistivity Si cells at room temperature and low illumination levels, and its efficacy in 10 ohm cm material at $100°C$ and 40 Suns is demonstrated in Table 5.1. The improvement in performance at high temperatures (relative to N^+/P cells) stems partly from the increased values of minority carrier diffusion length, which allow the photocarrier-confinement and dark current-suppression qualities of the BSF to be fully exploited. Furthermore, the contribution of the P/P^+ junction to the output voltage exhibits a positive temperature coefficient,[141] and so the degradation of V_{oc} with T is reduced. The alternative to a BSF arrangement is to use a lower resitivity base material, in which case high level injection effects would be reduced but degrading high doping density effects (Section 3.3.1) might occur. These two factors combine to suggest that for high temperature, high illumination operation a base resistivity of $0.3 \ \Omega$ cm is

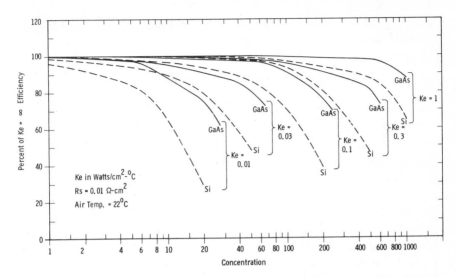

Fig. 5.2. Efficiency variation with sunlight concentration ratio for Si and GaAs cells as a function of heat sink thermal conductance. The ordinate represents the percentage of the ambient temperature $(22°C)$ conversion efficiency that can be obtained with a given heat sink. (After James et al.;[170] courtesy of IEEE)

optimum for N^+/P Si cells.[269]   It has been predicted that at $100°C$ and under 50 Suns illumination these cells could achieve an efficiency of 15%.

For GaAs cells the high value of bandgap and the dominance of recombination-generation mechanisms in the dark current ensure that V_{oc} is both higher and less degraded by increases in T than is the case for Si cells. Also, as minority carrier diffusion lengths are inherently short in GaAs, the improvements in J_{sc} brought about by increasing T are proportionately greater than in Si. The shift to lower energies of the absorption edge on increasing T can contribute to a further increase in J_{sc} but, conversely, the lowering of E_g for the window material in GaAs-based heterojunctions attenuates the high energy radiation reaching the junction region. The use of thin surface layers can minimize the latter effect and it is clear that the high temperature performance of GaAs cells is potentially superior to that of Si cells. This is illustrated in Fig. 5.2, from which it appears that for comparable performance at a given concentration ratio Si cells require a heat sink which is about three times as effective as that needed for GaAs cells.

5.2 GALLIUM ARSENIDE SOLAR CELLS

The basic structure of a high efficiency GaAs-based solar cell is shown in Fig. 5.3. The P on N rather than N on P arrangement is favored because of the availability of suitable dopants for forming the shallow homojunction.[103]   Doping with Zn, Ge and Be can be achieved by out-diffusion from the doped gallium aluminum arsenide, which is usually deposited by liquid phase epitaxial techniques.[126, 153, 170, 276]   Sequential liquid phase epitaxial formation of p-type GaAs and $Ga_{(1-x)}Al_x$As using Ge as a dopant is another possible method.[277]   Liquid phase epitaxy is preferable to vapor phase epitaxy on account of the fact that longer minority carrier diffusion lengths can be obtained in GaAs prepared in the former manner.[278] The n-GaAs layer is also usually grown by liquid phase epitaxy, either on n^+ substrates or on an n^+ layer which is in turn supported by an n-type substrate. The N^+/N junction is unlikely to provide any BSF action, but it has been found that an n^+ layer can serve as a useful

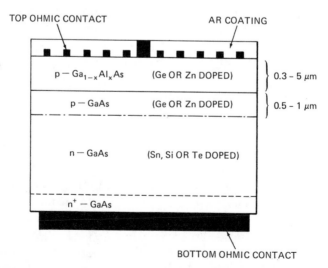

Fig. 5.3. Schematic cross-section of $Ga_{1-x}Al_x$As/GaAs high efficiency solar cell.

buffer between the main solar cell and the substrate, which may then be of relatively low quality (e.g., Te-doped rejects from the LED industry).[126] Cells of dimensions 2 cm X 2 cm and 1.27 cm diameter have been prepared by the above techniques, and such sizes should be adequate for operation at high concentration ratios.

To determine the thickness of the $Ga_{(1-x)}Al_x As$ surface layer the trade-off between decreased sheet resistance and increased photon absorption needs to be taken into account. A value of between 0.3–5 μm seems adequate, with the compositional factor x being about 0.8. Suitable ohmic contacts to the surface layer can be made with Pd, Au-Zn or Ag-Zn, while Sn or Au-Ge-Ni seem satisfactory for the back contact. Antireflection coatings that have been used with success are sprayed-on TiO_2[153] and Si_3N_4.[170] The latter film also shows some promise of being adequate as a protective coating for the solar cell, particularly if the device can be shielded from direct rain incidence. This is possible in the 1 kW array structure being fabricated by Varian Associates in the United States, as in this design the cells are inward-facing towards the concentrating mirrors.[51]

The high efficiency cells used in the latter array have already been discussed in Chapter 3. By using a contact grid pattern with 10% front surface area coverage the series resistance of 1.27 cm diameter cells can be made as low as 0.027 Ω. Whilst the long-term stability of these (and indeed of any) solar cells has not been completely established, some devices have operated for 7 months at illumination levels between 200 and 1000 Suns and temperatures between 20 and 95°C, with no signs of degradation.[51] The thermal performance of these cells is particularly good and even at 200°C an efficiency of 14% can be obtained with a concentration ratio of 312.[170] It has been estimated[279] that an array of such cells with a 10 MW peak rating would require only 80 m^2 of GaAs, and, as such cells need only be a few microns thick, there would appear to be no problem concerning the ability of world reserves of Ga and As to meet likely demands for photovoltaic power systems employing concentrators.

Even though solar cells are unlikely to be the most expensive item in high concentration ratio systems there is obviously some incentive to keep their cost as low as possible. In GaAs solar cells the cost is determined primarily by the price of Ga, which presently can be bought at around $1/gram. Because of this expense it has been estimated that operation of GaAs cells at concentration ratios greater than 500 Suns is necessary before any cost improvement over single-junction Si cells (which can be operated at around 100 Suns) can be realized.[126] In this respect it is noteworthy that GaAs cells have already been operated at 896 and 1735 Suns, yielding conversion efficiecies of 17.2 and 19.1% respectively (corresponding to individual cell outputs of 7.38 and 3.33 W respectively).[170] In any event, GaAs may well be compatible with lower cost fabrication procedures and thinner solar cells, as has been evinced by the production of ~20 μm thick cells by peeled-film techniques (Section 4.2.4 and Ref. 215). In this approach the single-crystal seed blocks can apparently be used "tens of times" and n-$Ga_{0.5}Al_{0.5}As/p$-GaAs devices with 13% efficiency have already been produced. Another development in high efficiency GaAs solar cell technology is the fabrication of AlAs/GaAs cells by vapor phase epitaxy techniques.[106, 128] The optical properties of $Ga_{(1-x)}Al_xAs$ improve as x approaches unity,[276] so AlAs is a logical material to try. As AlAs is hygroscopic encapsulation of the device is necessary; the anodically grown oxide may be suitable in this regard, as well as providing a useful antireflection coating. The AlAs/GaAs devices referred to in Chapter 3 have been operated at 200 Suns and yielded conversion efficiencies of 15.5%.[128]

5.3 SILICON SOLAR CELLS

It is not possible for silicon solar cells to match the performance of GaAs cells at very high temperatures (i.e., much in excess of 100°C), but operation under high illumination intensities is possible. Single junction devices appear to be useful up to at least 100 Suns, and operation at much higher concentration ratios can be realized using multijunction devices.

5.3.1 Single junction devices

Silicon homojunction solar cells, unlike $Ga_{(1-x)}Al_xAs$/GaAs cells, do not employ a heteroface structure with a high conductivity surface layer and thus must rely on metallization patterns with many fine fingers to reduce R_s to acceptable levels. Multiple terminal bars and supergrids that serve to divide a given cell surface into a number of smaller units are also often employed to further assist in reducing series resistance. In 2 cm X 2 cm cells, for example, 80 equally-spaced grid lines running between two edge terminal bars can give a series resistance of 0.034 Ω for a front surface coverage of 14%.[271] For a solar cell utilizing a full 5 cm diameter wafer with 17 fingers running perpendicular to, and bisected by, one tapered terminal strip $R_s = 0.4 \ \Omega$,[271] and thus such a cell could not be used at concentration ratios in excess of about 10 (Fig. 5.1). A much finer metallization pattern is required if such large area cells are to be used at high solar intensities. One design, for which operation at 100 Suns appears possible with negligible degradation in fill factor due to lateral resistance effects, uses a circular, peripheral terminal ring connecting to 240 fingers equally spaced in angular location.[269] The finger width is 0.0056 cm at the buss and tapers to 0.0013 cm near the center of the cell, giving a 10% metal coverage and effective cell area of 15.2 cm². Utilization of circular wafers without further cutting is an attractive proposition and the latter cells have been specifically designed for the 1 kW array using Fresnel lenses with $C = 60$ as described in Chapter 2.[27] Examples of top surface ohmic contact materials that have been used in Si concentrator cells are Cr-Au for p-type,[280] and 50 nm of Ti under several microns of Ag for n-type.[269] In processing the spokelike pattern described above the metal adhesion layer must first cover the whole surface prior to photolithographic defining. This temporary coverage of the active region with Ti has been found to cause deleterious surface effects and so Al is now used for this purpose.[269]

Because Si cells are appreciably cheaper than GaAs cells it is reasonable to consider designing the former for use at lower concentration levels than are practicable for GaAs devices, e.g., ranging from about 6 Suns for compound parabolic concentrators to about 100 Suns for Fresnel lens systems. The top contact metallization pattern would be one design element and another important one would be the base layer resistivity. Although high resistivity devices with back surface fields appear promising for concentrated sunlight applications there have not yet been any reports of successful employment of such devices. Present N^+/P concentrator cells thus use relatively low base resistivities and examples of devices that have realized high efficiencies at close to their concentration ratio design point are: ρ_b = 0.8 Ω cm, η = 12.5% at 17 Suns;[271] ρ_b = 0.38 Ω cm, η = 15.5% at 23 Suns;[271] and ρ_b = 0.3 Ω cm, η = 12.2% at 60 Suns.[269] The performances of these cells at concentration ratios other than the design point are encouraging; efficiency values of about 80% of the maximum value appear usual at concentration ratios four times higher than the design point.[271] Perhaps the most striking example is a 0.38 Ω cm cell, designed for use at about 20 Suns, which has yielded 12.8% conversion efficiency at 109 Suns.[271] All the above results refer to operation at temperatures around 20–30°C. At higher temperatures cell conversion efficiencies at the concentration ratio design point usually degrade by one percentage point every 16–22°C.[269, 281]

If higher resistivity base materials than quoted above are used high level injection effects can be expected at high sunlight intensities. Recent calculations suggest that under these circumstances it is preferable to employ P^+/N rather than N^+/P structures and so obtain a higher mobility majority carrier in the base region.[282] Some small area (0.16 cm $\times$ 0.16 cm) $P^+/N/N^+$ devices using $\sim$ 30 Ω cm base material have been constructed and operated at about 120 Suns to yield a conversion efficiency of 8%.[280] Some improvement in this figure may result from increasing the base layer thickness beyond its stated value of 50 μm, so enhancing the photon absorption in this particular device.

All the cells discussed thus far have been rectangular in crossection and as such would be mounted on a flat heat sink in the appropriate concentrator system. A departure from this geometry, in the form of a tubular p-n junction cell has already been mentioned (Section 2.2.2), and has some interest for use in solar concentrators of the cylindrical trough and compound parabolic varieties. The basic Si material is grown by edge-defined film-fed growth and a wrap-around inside contact facilitates the series connection of tubes.[52] Besides the matter of present low conversion efficiency (7%), the problems with this structure are the difficulties it presents in actual cell processing and in being positioned so as to be uniformly irradiated. The latter problem may also be present with planar cells used in nonimaging type concentrators, and could be serious as the resultant variation in electrical output from unit areas within the cell could cause a substantial decrease in total cell conversion efficiency.

5.3.2   Multijunction devices

Multijunction structures that have been investigated for Si solar cell use are shown in Fig. 5.4. The original vertical multijunction device shown in Fig. 5.4a has been discussed in detail[283, 284] and fabricated using selective etching techniques, with or without subsequent epitaxial refilling of the grooves.[285] The attraction of this device is its improved spectral response due to ubiquitous junctions, but the photogenerated current carriers still have to flow through high sheet resistivity regions before being collected, so series resistance problems are not alleviated. Furthermore, the effective dark current area is greater than the photocurrent area (as discussed in Section 3.3.2 for some Cu_2S/CdS cells) and so V_{oc} values for such devices could be low. However, if the N-type regions are allowed to penetrate the entire device thickness and current flow can be horizontally directed, then R_s can be made very low and a cell for use in high illumination environments is possible.

In one such design, 16 and 32 junction devices of the form shown in Fig. 5.4b have been constructed using 250 μm thick $P^+/N/N^+$ slices, bonded together at 700°C with 18 μm Al foils between slices.[286, 287] Subsequent scribing of the stack has produced cells with top surface areas of 0.16 cm^2 and 0.77 cm^2, thicknesses (from top to bottom in Fig. 5.4b) of up to 1 mm and conversion efficiencies of 6.2% at 200 Suns (temperature not specified) and 6.1% at 329 Suns (T = 130°C). Cooling of the devices would seem to be a first requirement for improving conversion efficiency, but some material design changes also appear necessary if concentration ratios much less than 1000 are to be used. This is because, although R_s decreases with P_i in these devices, values at 200 Suns are still around 3 Ω per junction. Low resistivity base material would improve this situation, although the possibility of any improvements from a high-low interface field (N/N^+ in Fig. 5.4b) would then be reduced. However, even in the present high resistivity devices (10–1000 Ω cm) the N/N^+ field is probably not being fully utilized as measured base layer minority carrier diffusion lengths are less than the base width.[287] Further improvements in the performance of this device should result from suitable passivation of the surfaces, application of a top surface antireflection coating and, perhaps, a bottom layer reflective coating. In one design a covering

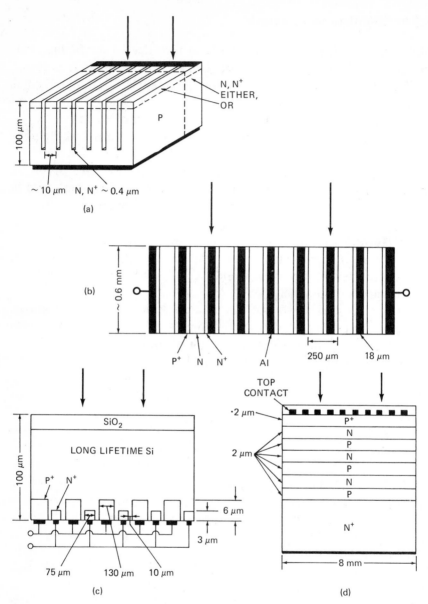

Fig. 5.4. Schematic diagrams of Si multijunction solar cell structures.

lens with a series of cylindrical refractive surfaces to concentrate sunlight onto each rectifying junction has been proposed.[288] With such an arrangement the minority carrier diffusion length requirements would be considerably relaxed, so allowing the use of lower resistivity material and the attainment of low series resistance.

A particularly effective design of Si concentrator cell is the interdigitated structure shown in Fig. 5.4c.[289] With this arrangement no top surface shading is present and carrier separation occurs near the back surface. A copious supply of electrons and holes in that region is assured by the use of high lifetime base material and a transparent SiO_2 top covering to reduce surface recombination losses. The use of high resistivity base material allows conductivity modulation to occur and a value of $R_s = 0.022$ Ω has been recorded at high sunlight intensities. As with most Si cell designs, maintenance of a low device temperature is necessary for good performance, and at $15°C$ and 220 Suns a conversion efficiency of 16.5% has been obtained. This figure is the highest yet reported for Si concentrator cells. An additional antireflection coating and the use of thicker back surface metallization could lead to even higher efficiencies in this device.

A different concept of concentrator cell design is embodied in the horizontal multijunction structure shown in Fig. 5.4d.[290] In this device the internal junctions serve to guide photogenerated carriers in the direction desired for good photovoltaic performance, i.e., holes toward the p^+ surface layer and electrons toward the n^+ base. As in more conventional multilayer electronic devices (e.g., Shockley diodes and SCR's) the accumulation of majority carriers in the various regions of the device forward biasses the internal junctions and aids the required flow of minority carriers. Provided the internal layers are thinner than the minority carrier diffusion lengths the collection efficiency will be high. By using 2 μm thicknesses it has proved possible to use p and n layers of ~ 0.1 Ω cm material and a six layer structure has produced a conversion efficiency of 11.1% at 623 Suns ($T = 56°C$).[290] Even though conductivity modulation is unlikely in this device the low resistivity of the layers and a suitable top contact metallization pattern (20% coverage) have allowed attainment of R_s values as low as 0.012 Ω. The output power density, relative to total front surface area, of this device was 5.5 W cm^{-2} and is the highest value reported for a Si solar cell. The corresponding value for a GaAs cell operated at 896 Suns (assuming 80% front surface active area) is 10.4 W cm^{-2}.[170] Thus, even in the realm of very high concentration ratio applications, where GaAs cells are often considered to be unchallenged, Si cells may in fact prove useful.

6.

Economic Assessment of Photovoltaic Power Systems

Solar-electric power plants are popularly envisioned as operating independently of nonsolar generators of electricity, and supplying energy in bountiful amounts without pollution of the environment. Huge satellite solar power stations of the type discussed in Section 2.6 may one day come close to realizing this ideal, but any terrestrially-located photovoltaic systems capable of operating in an independent manner are likely to be small ($\lesssim 10$ kW). Nevertheless there are many applications for which plants of this size and type are appropriate (see Table 1.2 and Section 2.5). In these instances the cost of photovoltaic electricity is quite simple to calculate, being just a matter of dividing the capital costs (due to the array plus power conditioner plus energy storage facility) by the rated output power, and then using the resulting $/W figure in conjunction with the estimated system lifetime to arrive at the cost per kilowatt-hour. There are no fuel costs and the expenditures associated with land purchase, transportation to the site, assembly and operation and maintenance are likely to be negligible for small systems. The lifetime of the system thus sets a maximum period over which the initial capital costs might be amortized. As the system size increases both capital and noncapital costs rise; e.g., costs due to support structure and intermodule connection must be allowed for in the former and interest, insurance and maintenance costs become important in the latter. Also, reliability is not associated solely with likely component failure but is linked to the variable nature of the insolation. It is no longer sufficient to size the power plant on the basis of worst-case conditions as this would likely lead to the need for enormous storage facilities (charged from the array) in order to safeguard supplies during periods of inclement weather, darkness, and array repair.

When considering loads typical of residential, commercial and industrial consumers in developed countries, the only situations where stand-alone operation of photovoltaic plant is likely to be possible are in modest, energy-conservative houses in high insolation areas and utilizing either individual or communal facilities for energy storage. In all other instances some interconnection with nonsolar generating plant will be necessary to maintain existing standards of electricity supply. A dedicated diesel generator, for example, might allow some residential and light industrial concerns to obtain the supplementary power they need but, in the great majority of cases, interconnection with the existing electric utility will no doubt be sought. Assessing the cost of electricity under these circumstances then becomes quite complex.

With proprietor ownership of the solar array one of the important factors to be taken into account in assessing the true worth of a photovoltaic system is the rate that might be charged by the utility for having to maintain services to an area of greatly reduced load factor. Under present rate structures the reduction in energy taken from the utility would not be matched by a comparable monetary saving for the consumer. Another concern of the owner is likely to be the extent of the rebate that might be granted by the utility for being able to use surplus photovoltaic power fed into its electrical network. From the utility standpoint the question would be whether the fed-back power from many privately-owned arrays could constitute a sufficiently reliable and manageable input, so that some displacement of conventional fuel or generating capacity might be possible. Realization of this possibility would be enhanced by collection of the fed-back power in a utility-owned, central storage facility.

Some of these possible problems at the private owner/utility interface could be circumvented by a situation in which the utility was responsible for the purchase, installation and operation of the entire photovoltaic system, irrespective of where the various elements were located. This option may well materialize because the capital-intensive nature of photovoltaic systems could be a considerable deterrent to possible purchasers in the private sector. Another factor in favor of this alternative is that the modularity of photovoltaic arrays, whilst being a constructional attraction, does not allow particularly large savings to result *a priori* from building larger and larger individual systems. Residences, commercial buildings and industrial areas, as well as more open country for central stations, could then all be regarded as potential locations for photovoltaic arrays capable of contributing to supplying the electrical needs of a region. In this arrangement photovoltaics would become just another generation method, albeit with very positive and salient features of renewability, availability and cleanliness, that could be accommodated into the mix used to meet the variety of present-day demands. The costing of photovoltaic electricity could then follow existing utility practice for new plant, with the only novel features being zero fuel costs and fluctuating supply. Under these circumstances the question of energy storage could then be addressed separately and its inclusion into the system decided upon following the present practice (Section 2.4).

The result of economic analyses for any of the above scenarios would be an energy cost figure (in mill/kWh), of which two of the main determinants would be the cost and lifetime of the solar array. Much of this book has been concerned with the factors affecting these two properties and the prospects of attaining the respective goals of 50¢/W and 20–30 years appear promising. The cost of arrays is not solely a question of technology, however, but is in fact influenced by a complex web of interactions. Amongst these can be counted various institutional and manufacturing factors (Section 2.7), as well as considerations of the availability and social desirability of other competing energy sources. The situation is summarized in Fig. 6.1, and at the present time it is difficult to gauge the pace at which the development of terrestrial photovoltaics will proceed. What is clear is that there is an increasing demand for electricity; that there are problems of renewability, supply and pollution with many existing fuels used in electricity generation; and that photovoltaics has considerable merits as a useful, large-scale source of electricity.

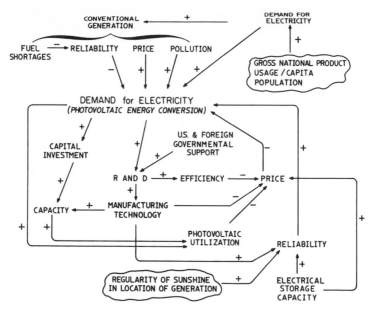

Fig. 6.1. Schematic representation of the factors which affect the development of photovoltaic power systems. An increase in the level of a cause (tail of arrow) results in an increase (+) or decrease (-) in the level of an effect (point of arrow). (After Merrigan;[9] courtesy of MIT Press)

The chain of developmental events illustrated in Fig. 6.1 is already being forged and in the United States, where activity is greatest, it seems likely that in the period 1985–2000 substantial penetration of photovoltaics into the electrical-supply market place will occur.[9, 291, 292]

One of the first questions to be answered once the decision to construct a particular photovoltaic power system has been taken, will be whether or not to use sunlight concentration. The influence this has on array design was discussed in Section 2.2 and the solar cells that might be used in various arrays were described in Chapters 4 and 5. So far the treatment has been mainly from a technical point of view; in the next section the economic aspects of array selection are addressed. Following this, assessments of the price that will probably have to be paid for photovoltaic electricity at various levels of consumption and under various conditions of ownership are presented.

6.1 ARRAY OUTPUT FIXED COSTS

The capital costs associated with an array can be broken down into three items, namely: module costs e.g., the solar cells and their interconnection, encapsulation and mounting on a suitable supporting heat sink; land costs; and various other expenditures due to some or all of the following—transportation, connection of modules, construction of supporting structure, concentrators, tracking apparatus, forced cooling equipment. A convenient expression for the total output fixed costs of an installed array is[293]

$$Q = (X/C + Y + Z/P_f)(fcr)/\xi \qquad (6.1)$$

where Q is the fixed cost in dollars per kilowatt hour (electrical), C is the sunlight concentration ratio, X is the module cost per unit area of module, Z is the land cost per unit area of land, P_f is the land utilization factor, Y is the other capital costs expressed on the basis of dollars per area of collector (aperture), ξ is the electrical energy generated per year per collector area and fcr is the fixed charge rate. The latter term accounts for depreciation, the cost of money, insurance and taxes and is used so that Q can be expressed in a levelized form, i.e., a single figure that is fixed at a constant value for the lifetime of the system and is so chosen to give the same end-of-life total costs as would obtain in the practical situation where the costs are likely to increase with time. Thus the levelized charge represents an overcharge (relative to increasing costs) in early years and an undercharge in later years.[294]

The above approach to array output fixed cost accounting has been used for flat plate arrays and various concentrator arrangements, under a variety of cooling conditions and at both high insolation (Albuquerque, New Mexico) and low insolation (Cleveland, Ohio) locations.[293, 295] Different costs are associated with different arrays and, of course, the collected radiation will depend on the array orientation and location (see Fig. 2.5). This means that it is unlikely that there will be an array design that is universally optimum. An example of the results obtained from Equation 6.1 for the case of a particular concentrator (2-D tracked, point focus concentrator with passive cooling) at a particular location (Albuquerque, 1962 weather

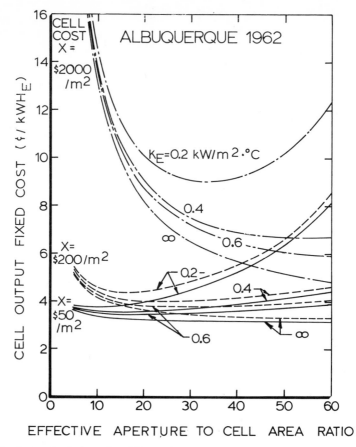

Fig. 6.2. Cost calculations for 2-D tracked, point focus concentrators with passive cooling: Si cells with $\eta = 15\%$ at $25°C$; $Y = \$60\ m^{-2}$; $P_f = 0.35$; $Z = \$1\ m^{-2}$; based on insolation data for 1962. (After Evans et al.;[293] courtesy of Pergamon Press)

data) is shown in Fig. 6.2. For modules of 15% efficient cells and an internal packing density of 0.9 the X values quoted span the solar cell cost range of \$15/W to about 40¢/W, i.e., from present to projected levels for silicon. Lower levelized energy costs result from improved heat-sinking because of the concomitant increase in cell conversion efficiency. However further increasing the effective thermal conductance K_E of the array by forced cooling techniques may not necessarily lower Q because of the associated increase in structural and cooling equipment costs. The latter are accounted for via the factor Y in Equation 6.1, but precise specification of this term is difficult at the moment because of the dearth of practical systems. A figure of $Y = \$60 \text{ m}^{-2}$ has been used for Fig. 6.2,[293] but for the experimental 2-D tracked Fresnel lens system ($C = 60$) discussed in Section 2.2.2 it is possible that a figure nearer $\$200 \text{ m}^{-2}$ might be appropriate.[27] For realistic values of K_E the cost curves exhibit minima, i.e., optimum values of C. Initially Q falls with increasing concentration ratio due to the decreasing contribution of the first term in Equation 6.1 (X/C) to the capital costs, but at higher values of C the increase in cell temperature causes the conversion efficiency and hence array output energy ξ to decrease.

It might be expected that, should solar cell array costs ever drop to $\$50 \text{ m}^{-2}$, it would not be profitable to employ concentrator arrays. However, data given in Ref. 293 suggest that this may not be so in all cases, e.g., a south-facing, flat plate, silicon array located at Albuquerque, tilted at the local latitude and with $P_f = 0.6$ and $K_E = 0.03$ kW m^{-2} °C^{-1} would yield an array output fixed cost of 3.4 ¢/kWh. The less expensive structure of the flat plate array (Y = \$15 m^{-2} in this case) is balanced by a reduced electrical output, so that costs for the two systems are similar. Even in low insolation areas with frequent cloud cover, e.g., Cleveland, Ohio, output costs for south-facing flat arrays tilted at the local latitude are predicted to be about the same as for 2-D tracked concentrators.[295] This cost floor is about 6¢/kWh for Cleveland, as opposed to about 3.5¢/kWh for Albuquerque, indicating the difference in insolation between the two places. In practice at $X = \$50 \text{ m}^{-2}$ flat plate arrays would probably be preferred to concentrator arrays owing to the cheaper noncapital costs arising from simpler operation and maintenance. Cadmium sulfide (Cu$_2$S/CdS) cells are potential candidates for operation in unconcentrated sunlight and it has been estimated that volume production using chemical spray techniques (Section 4.3.1) could yield cells for as little as 6¢/W.[184] This very low figure would give an X-value of $\$2.7 \text{ m}^{-2}$, assuming an internal packing density of 0.9 and the present value for the conversion efficiency of these cells i.e., 5%. Because of this low efficiency large support structures and many interconnects would be needed to fabricate an array and it has been estimated that costs associated with these functions would be about $\$17 \text{ m}^{-2}$.[172]

The poor elevated temperature performance of CdS/Cu_2S cells would probably demand that forced cooling be used, so that a Y value of about \$40 m^{-2} may not be unreasonable. To compare such cells with the Si arrays used to compute Fig. 6.2 take the above values of X, Y and η and assume, optimistically, that all other salient parameters are the same as for Si arrays with $K_E = 0.03$ kWm^{-2} $^{\circ}C^{-1}$. Under these conditions for an unconcentrated array at Albuquerque the array output fixed costs would be 7.9¢/kWh.

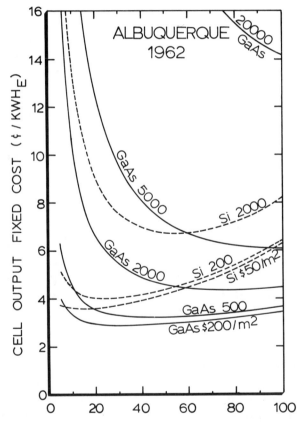

Fig. 6.3. Cost calculations for Si and GaAs cells using 2-D tracked, point focus concentrators with passive cooling. In all cases $K_E = 0.4$ $kWm^{-2}$$^{\circ}C^{-1}$; otherwise Si data same as in Fig. 6.2; GaAs cell $\eta = 20\%$ at $25^{\circ}C$. (After Evans et al.[295] with permission)

An alternative material to silicon for use in concentrator arrays is gallium arsenide. The higher conversion efficiency and better high temperature performance of this material (Chapter 5) are attractive features, but the cost of cells is probably at least an order of magnitude greater than for silicon, although there are no commercial products available at present to confirm this. At these large values of X higher levels of sunlight concentration are required to realize the benefits of GaAs cells (see Fig. 6.3), although if the cells could be produced for sale at $200 m^{-2} the increased conversion efficiency (assumed to be 20% in this case) would render GaAs superior to Si even at very low values of C.

6.2   THE COST OF ELECTRICITY FROM PHOTOVOLTAIC POWER SYSTEMS

In order to compute the cost of electricity from a photovoltaic power system capable of supplying loads typical of present-day residential, commercial and industrial consumers a variety of charges need to be added to the array output fixed costs discussed above. Amongst these are capital cost charges due to power conditioning, energy storage, nonsolar electricity generating plant, and running costs due to operation and maintenance and the use of some nonsolar fuel. The question of ownership of the array also has a bearing on the accounting structure. At the residential (1–10 kW) and intermediate (0.1–10 MW) power levels both proprietor and utility ownership might be reasonable, but at the central station level (50–5000 MW) utility ownership can be assumed. Examples of cost calculations for these various circumstances are presented below.

6.2.1   Residential systems

The economic viability of a proprietor-owned, on-site photovoltaic power system is dependent upon the cost of the system and the value to the owner of the energy savings, in the form of reduced purchases from the utility, that the system allows. As a first economic analysis consider a system comprising a solar cell array, battery storage facility and a utility tie-in for supplying power only in the event of both the energy-store being exhausted and the insolation level being insufficient to meet the load requirements. In this case utility power is always on-hand, the questions of fed-back power and utility charging of the energy store do not arise, and the savings in energy can be relatively simply calculated, assuming particular insolation and load patterns. The fraction of the load supplied by the photovoltaic power plant will depend upon the area of the array and the capacity of the energy store. There is likely to be an optimum sizing arrangement as too big a storage capacity for a given array area would mean that the store was never fully charged, and too large an array area for a given storage capacity would mean that photovoltaic electricity was often wasted (if there was no provision for feedback to the utility).[54,296]

Once the units' sizing is decided upon the energy savings from the utility can be predicted, and a convenient form of expression for the "worth of the system" is via a figure of merit F_M, defined as the system cost (in dollars) divided by the annual reduction in energy purchased from the utility (in kWh).[54] For example, the F_M values for the residential rooftop arrays in Phoenix and Cleveland that were described in Section 2.1 can be estimated as 0.61 and 0.90 respectively. These figures were arrived at by dividing the calculated photovoltaic output energies (17,461 kWh for Phoenix and 11,901 kWh for Cleveland) into an assumed system cost of $10,700 i.e., the residence roof area multiplied by $100 m^{-2}. The latter figure is intended to represent the installed cost of all components plus the present value of all future repairs, maintenance and replacements likely to be incurred over a thirty-year lifetime. Assuming a Si array cost of $50 m^{-2} this leaves a like sum for the operational costs, which seems a reasonable estimate for the above examples as no energy storage facilities are employed. To see under what circumstances the above F_M values would indicate the cost effectiveness of installing a photovoltaic system consider Fig. 6.4, in which F_M is plotted against the utility charge rate with the effective interest rate i as a parameter. The latter represents the difference between the mortgage interest rate and the escalation rate for fuel as used by the utility. The mortgage rate enters into the calculations as it is assumed that payment for the photovoltaic system will be financed via the mortgage for the residence. Values of i between 8% and 4% have been suggested as being appropriate for the United States for the period up to 2000,[54] thus at the median value of $i = 0.06$ the F_M values computed above would indicate cost effectiveness of the photovoltaic system installations if the electricity rate was greater than about 4.5 ¢/kWh in Phoenix and 6.5 ¢/kWh in Cleveland. Judging by the cross-hatched areas shown in Fig. 6.4, which cover likely future rates and F_M values for photovoltaic systems installed across the United States it would appear that the system in Phoenix may be economically viable in 1985 and that in Cleveland by 2000. Photovoltaic systems are likely

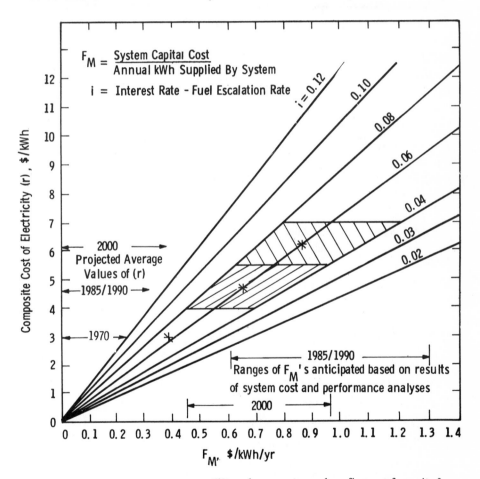

Fig. 6.4. Relationship between utility charge rate and a figure of merit for proprietor-owned photovoltaic systems (see text), for various values of composite interest rate. (After Federmann et al.;[54] courtesy of IEEE).

to increase in attractiveness for all locations as time progresses owing to system cost reductions due to volume production and higher utility electric rates due to fuel shortages.

A different accounting procedure is appropriate for utility ownership of residential systems and it is convenient to express the relevant energy costs (in constant year dollars per kWh) in a levelized fashion. For rooftop systems similar to those discussed above the results of an analysis using economic and financial factors germane to a utility-owned enterprise (Table 6.1) are shown in Fig. 6.5.[29,297] The dashed curves represent costs for a system with no energy store but allowed feedback of power to the utility, whilst the solid curves refer to a

Table 6.1. Economic Assumptions for the Analysis of Utility-owned Residential Photovoltaic Power Systems as Used to Compute Fig. 6.5.[a]

Capital structure
    Debt fraction   : 50%
    Equity fraction : 50%
    Risk premium   : 2.5% (Difference between interest on debt and return on equity)

Before tax discount rate
    Varied parametrically from 5 to 15 percent

Escalation Rates:

	1975–1985	1986–2019
General price level	3.8	4.2
O&M labor	4.8	4.5
Materials	3.1	3.8
Installed capital	4.8	4.6
Insurance	4.8	4.6

Federal tax rate: 48%

Other taxes: None assumed.

Year of capital expenditure: 1989

Start of operation: 1990

System lifetime: 30 Years

Depreciation method: Sum of years digits

Investment tax credit: 10 percent

[a]From Ref. 297.

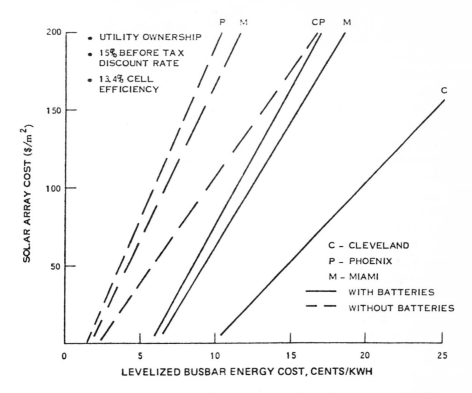

Fig. 6.5. Levelized energy costs for residential photovoltaic systems (utility-owned). The dashed lines are for the case of no energy storage but allowed feedback of power to the utility; the solid lines are for the case of some energy storage but no allowed feedback of power. The economic assumptions made in the analysis are given in Table 6.1. (After Kirpich et al.;[29] courtesy of IEEE)

photovoltaic system with battery storage (at $77/kWh) and no provision for feedback of surplus power to the utility. To produce photovoltaic power at a levelized cost of 5¢/kWh would apparently require array costs to be about $75 m^{-2} for use in Phoenix and $30 m^{-2} for use in Cleveland. In the earlier discussion an array cost of $50 m^{-2} was assumed and using this figure in Fig. 6.5 suggests electricity costs of about 4¢/kWh and 6¢/kWh for Phoenix and Cleveland respectively. It thus appears that the cost of photovoltaic electricity is very similar irrespective of whether the system is owned by the proprietor or the utility.

Figure 6.5 suggests that the use of battery storage may not be cost-effective in some photovoltaic power systems and the principal reason for this, perhaps unexpected, result is that the energy consumed in charging the store leads to a significant reduction in energy displacement. However, in the above scheme where no battery storage was used, the system effectively employed the utility network as its storage facility and thus, in the event of large price increases in conventional fuels, this approach may become less attractive. Also, with the increased penetration of photovoltaic power systems into the electrical grid network it may prove necessary to exert some control over the potentially large amounts of fed-back power, and this might be most expeditiously achieved by using this power to charge an energy storage facility. With such an arrangement residential rooftops would become dispersed generating sites acting, in effect, to provide power for an entire area and not just the co-located houses. The worth to the utility of such a scheme could be assessed by comparing the costs due to the arrays, energy store and backup, nonsolar power facility with the cost of supplying the region entirely by conventionally-generated power. Such an analysis has been carried out for the central Arizona region where the assumed new load to be supplied consisted of 35,000 houses consuming a total average power of 121.8 MW and a total peak power of 287 MW.[296] To supply this load from a conventional power station connected into the existing grid would require 267 MW of new plant, and this would have to be located some distance away (200 miles was assumed)[296] from the load center so transmission costs must also be considered. On the other hand, if each dwelling used its south-facing roof (114 m^2) for a photovoltaic array a total array area of 4×10^6 m^2 would be available which, when combined with a central energy storage facility of capacity 10^6 kWh (28.5 kWh per residence or 8.2 h supply at average demand), would mean that only 73.3 MW of conventional backup capacity would be required.[296] Transmission costs in the latter case (computed on a dollar per MW-mile basis) would thus be reduced by about 4 times, as would be capital and operating costs for the conventional facility (assumed here to be a coal-fired plant constructed at \$550/kW). The required break-even array price (i.e., the price of photovoltaic arrays that would equalize the present value costs of the two schemes) is shown in Fig. 6.6 for various values of conven-

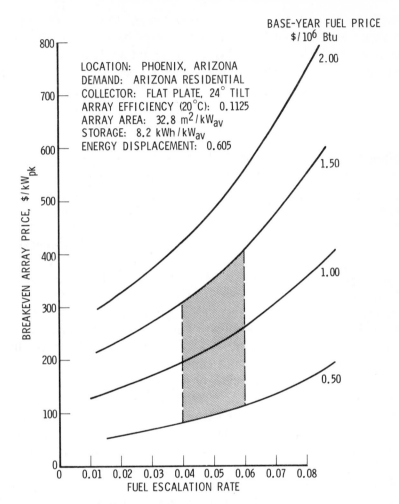

Fig. 6.6. Breakeven array price, for residential photovoltaic systems, in terms of
various alternative fuel-price projections. (After Leonard;[296] courtesy of
IEEE and National Science Foundation).

tional fuel price. The shaded area represents the ranges of values for
1990 fuel costs and post-1990 fuel escalation rates that appear to be
reasonable.[296]   As would be expected, the break-even array price is
very sensitive to conventional fuel price and, for the example cited
above, values in the range of $100/kW_{pk} to $400/kW_{pk} are predicted. These values are just below the 50¢/W figure frequently

quoted as the goal for solar cell array prices. Cost-effectiveness at this figure and a 6% fuel escalation rate could be achieved if the base year price (in 1976 dollars) of coal was about $1.75/10^6$ Btu. Some North American utilities are already paying close to this price as the range across the continent runs from about 40¢ to $1.50 per 10^6 Btu.*

6.2.2 Intermediate-level systems

Analyses of intermediate-level (100 kW to 10 MW) photovoltaic power systems cannot be performed in such a general manner as is possible for residential systems as the nature of the probable load is not so well-defined, e.g., the consumer could be an industrial or commercial concern or an entire utility distribution substation. Because of the collector area required to supply power at the megawatt level (10^4 m^2/MW_{pk} for 15% efficient cells and an overall array packing factor of 0.67) it is unlikely that such large units could be accommodated close to existing residential areas. However in the cases of commercial and industrial premises there may be sufficient roof area and adjoining space (over parking lots and warehouses for example) upon which a sizable photovoltaic array could be installed. Shopping centers, large administration buildings and some light industrial parks would be particularly suitable as the demand for electricity would be mainly during daylight hours, and so the photovoltaic power could be used directly and storage requirements greatly reduced. Furthermore, lighting, air-conditioning and various industrial processes that might dominate the loads at some of these sites may well be amenable to d.c. supply, thus obviating the need for expensive inverters ($\sim$$275/kW).[45] The tie-in to the utility could then supply any a.c. loads and meet any surplus d.c. requirements via a cheaper rectifier unit ($\sim$$55/kW),[45] perhaps conveniently using an energy storage facility as a buffer. The question of back-feeding power to the utility would not normally arise because the available collector area would not be expected to be capable of generating electricity in excess of the demand, except, of course, on holidays.

---

*By way of comparison oil ana natural gas cost closer to $2/10^6$ Btu and the effective price of uranium is about 10¢/10^6 Btu. These latter figures cannot be used directly in Fig. 6.6, however, as capital costs for plants using these fuels are not the same as for coal-fired plants.

Photovoltaic power generated at these times could be stored, and the incorporation of an energy storage facility may also be sensible on the grounds that it could be charged during any working day by some of the electricity displaced from the conventional generating facility. Provided charging in this manner did not lead to an increase in the peak load supplied by the utility, the operation would be cost-effective (from the viewpoint of a proprietor-owner) as present rate structures to intermediate-load consumers tend to be on the demand-plus-use basis, i.e., a substantial charge is placed on the level of peak power drawn by the consumer, and a relatively small charge on the amount of energy used.[54]   In both proprietor-owned and utility-owned systems the strategy for energy usage is likely to be one in which the solar array serves an energy displacement function, feeding the load whenever power is being generated, whilst the battery store is used for load levelling or peak shaving.

A specific example of an intermediate-level photovoltaic power system has recently been proposed in the form of a supply for a shopping center in Los Angeles, California.[45] The typical daily electrical consumption for a center of the proposed size (23,226 m² floorspace) is about 37,000 kWh, but by judicious use of d.c. lighting and using d.c. for the main load of the air-conditioning plant, the design consumption level was able to be set at about 14,000 kWh/day.[45]

As load matching for this example is particularly good (in typical shopping centers in Los Angeles 70% of the load occurs in daylight hours and peak loads are in summer) a storage facility was not included in the proposed power system and, even without it, the photovoltaic power unit was predicted to be able to provide 45% of the annual load (i.e., the plant capacity factor was 0.45). The solar cell modules were assumed to be arranged in a sawtooth arrangement (Fig. 2.10b) and a breakdown of the module costs is given in Table 6.2. For 0.91 m X 2.44 m modules composed of Si cells rated at 50¢/W the purchase price was computed to be $64 m⁻². The total photovoltaic system cost was estimated to be $953,000 and the annual solar energy delivered to the load was 2,090,000 kWh. Thus assuming, perhaps pessimistically, that maintenance and replacement costs over a 30 year period would equal the original purchase and installation cost leads to a F_M value of 0.91. At a composite interest

**Table 6.2. Summary of Module Fixed Costs for 0.91 m X 2.44 m
Unit of Silicon Solar Cells.[a]**

Materials	kg/Module	$/Module
Glass, 1/16" low iron	22.5	$ 6.24
Polyvinyl butyral interlayer	0.1	4.01
Aluminum frame and mounting hardware	5.9	9.19
Silicone rubber seal strip	0.1	4.50
Electrical connectors	0.2	1.80
	28.8	$25.74
Labor (At $12/hr including overhead)	Hrs/Module	$/Module
Cell interconnection and test	1.00	$12.00
Laminate and assemble to frame	0.75	9.00
Add terminations, final test & inspection	0.25	3.00
Packing for shipping	0.05	0.60
	2.05	$24.60
Manufacturing cost (excl. cells)		$50.34
Distribution expense (15%)		7.55
Profit (20%)		10.07
Selling price per module (excl. cells)		$67.96
per square meter		$30.48
Elec. Output (15% cells, 0.74 packing, std. test cond.)		247.5 watts
Allowable cost at $500/kW		$123.75
Allowable cell cost		$ 55.79
per square meter of cells		$ 33.81
per kW—cells only		$225.40

[a]From Ref. 45.

rate of i = 4% this suggests, on the basis of Fig. 6.4, a cost-effectiveness of the proposed power system should the utility charge rate exceed 5 ¢/kWh.

The economic viability of the above system thus seems to be dependent upon attainment of array price levels and utility charge rates of very similar magnitudes to those required for implementation of residential photovoltaic power systems.

6.2.3 Central power station systems

A photovoltaic central power system rated at the 1000 MW level, for example, would occupy about 10^7 m² so the question of its location is not trivial. Deserts, mountainsides, sheltered waterways (including existing dams) and other suitably insolated open spaces are all possibilities. Siting away from relatively congested urban and industrial areas would allow the consideration of more elaborate tracking methods, and hence arrays with higher concentration ratios, than are feasible for residential and intermediate-level systems. A photovoltaic central power plant would doubtless constitute an integral part of a region's electricity supply network and be coupled into the grid via transformers. The latter would be necessary to increase the voltage from its source value of around 1 kV (the probable limit set by array isolation and semiconductor inverter limitations), to the prime or intermediate voltage levels which would be desirable because of the photovoltaic plant's size and remote location.[54]

There would appear to be three principal configurations in which the photovoltaic central power station might operate, namely: (1) in a stand-alone fashion utilizing an enormous energy storage facility; (2) in conjunction with a smaller energy store and a backup, nonsolar power plant; (3) in conjunction with nonsolar power plant but with no form of storage. In the first arrangement the energy store would be charged by the array whenever the load conditions and insolation permitted, and the sizing would have to be such that the store was always sufficiently charged to be able to meet the load demand during the night and in periods of poor weather. The required magnitude of storage capacity in this application could probably be met by a variety of methods (see Section 2.4), provided the required sites, technologies and funds were available. Even when this time is reached, however, there will still be one undesirable feature of this approach, namely the charging of the energy store by a generation method with zero incremental generation costs. Because of this latter property, and the fact that the solar day corresponds to a large extent with the intermediate load period (i.e., between the morning and evening peak demands), it seems natural that the photovoltaic central power plant be associated with intermediate-load generation and be dispatched whenever available, rather than attempting a system design which would allow base-load operation. Operation in an intermediate load mode is the basic tenet of the second and third approaches listed above.

In case (3) energy storage is not used at all, and the photovoltaic

Table 6.3. Summary of the descriptions (a), characteristics (b), and estimated costs (c) of photovoltaic central power stations as designed for Phoenix, Arizona.[a]

(a)

	PARABOLIC PONTOON SYSTEM	TWO ELEMENT CONCENTRATOR	TRACKING CPC SYSTEM	SEASONAL ADJUST CONCENTRATOR	REFLECTOR AUGMENTED FLAT ARRAY	FLAT ARRAY
	I	II	III	IV	V	VI
Analyzed by	GE	Spectrolab	Westinghouse	GE	Westinghouse	GE
Geometric concentration ratio	25:1	20:1	10:1	3.6:1	1.5:1	1:1
Concentrator type	Parabolic trough	Parabolic trough/ compound elliptic concentrator	Compound parabolic concentrator	Parabolic half-trough	Array/reflector sawtooth	None
Tracking	Azimuth	Azimuth	Azimuth	Seasonally adjusted tilt	Fixed	Fixed
Orientation/Tilt	Horizontal	45° tilt	50° tilt	EW, horizontal	53° - array 30° - reflector	22°
Cell type	Silicon	Silicon	Silicon	Silicon	Silicon/CdS	Silicon
Cell standard efficiency	11.8%	15%	16%	13.4%	16%/10%	13.4%
Co-located storage	None	3-hour Lithium-sulfur	None	None	None	None

Table 6.3 (*continued*)

(b)

DESIGN	I-GE	II-S	III-w	IV-GE	V-w	VI-GE
Nominal plant rating	1500 MW	200 MW	100 MW	1500 MW	100 MW	1500 MW
Required land area, km^2	50	11	5	39		24
Annual energy output, MWh/yr	2.83×10^6	0.58×10^6	0.27×10^6	2.83×10^6	0.18×10^6*	2.83×10^6
Collector area, km^2	25	2.08	0.95	32.6		12.9
Cell area, km^2	1	0.104	0.095	7.3		10.1
Inverter module size, MW	34	2	4.4	34	4.4	34
Inverter input voltage (array voltage), vdc	13000 (±6500)	1200 (±600)	1000 (1000)	13000 (±6500)	1000 (1000)	13000 (±6500)
Inverter output voltage, kvac	69	34	6.9-34.5	69	6.9-34.5	69

*For silicon cells

Table 6.3 (continued)

(c)

DESIGN	I-GE	II-S	III-W	IV-GE	V-W	VI-GE
System cost 1975 $/kW$_p$	930	1445[1] 2025[2]	1046	1218	775–810	875
Cell cost, $/m^2-Cell $/kW$_p$ @ 1 SUN	50 425	182[3] 1210	88 500	50 425	88 500	50 425
Cell cost, $/kW$_p$ $/m^2-Array	33 2	95 9.1	84 8.8	243 11.2	220 59	337 39
Array cost (w/o cells) $/kW$_p$ $/m^2-Array	600 36	320 30.7	260 27.2	526 24.2	70–79	202 23.5
Structure cost $/kW$_p$ $/m^2-Array	89 5.3	650 62.5	274 29	231 10.7	144–160	154 17.9
Inverter cost $/kW$_p$ $/m^2-Array	95 5.7	142 13.6	108 11.4	95 4.4	108	95 11.1
Misc. Cost $/kW$_p$ $/m^2-Array	113 6.7	238 22.9	320 33.7	123 5.6	233–243	87 10.1

[a]From Ref. 298.
*Estimated for large volume production.
1) No Storage   2) 3-hr Storage   3) Remainder of column relates to no-storage system.

power plant would deliver energy to the system load throughout the hours of daylight. The benefit to the utility of this approach would be largely related to the reduction in fuel consumption that would be allowed at other, nonsolar power stations within the electrical supply network. This scheme has been favored by the majority of investigators in the United States who recently submitted to ERDA their designs for a photovoltaic central power station to be located in Phoenix, Arizona, see Table 6.3. In all cases levelized energy costs in the range of 30–70 mill/kWh were predicted for array costs of $500 per peak kilowatt and array (not just cell) conversion efficiencies of 10% or greater.[298] Some results from the General Electric Company's designs (cases I, IV and VI in Table 6.3) for a 1500 MW plant are shown in Fig. 6.7. At a cell cost of 50¢/W and conversion efficiency of 12.5% cases I and VI yield electricity at a levelized cost of 4-5 ¢/kWh, i.e., the same cost as found applicable to the no-storage, roof-top residential array system as discussed in Section

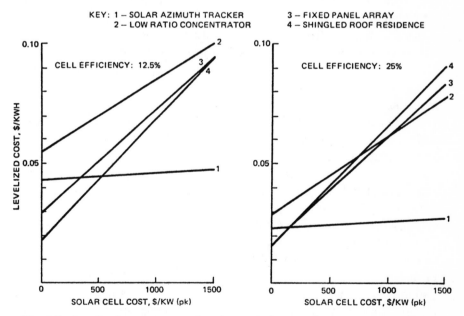

Fig. 6.7. Levelized energy costs for photovoltaic central power systems. Curves 1, 2, 3 refer to cases I, IV, VI, respectively, in Table 6.3. Curve 4 is for the residential system discussed in Section 6.2.1. Location is Phoenix, Arizona. (After Kirpich et al.;[29] courtesy of IEEE)

6.2.1 and included in Fig. 6.7 as curve 4. The parabolic concentrator system mounted on floating pontoons (case I) seems particularly attractive and even more so at high solar cell conversion efficiencies. The doubling of cell efficiency shown in Fig. 6.7 leads to almost a halving in subsequent system energy costs and even at $1.50/W levelized costs are predicted to be as low as 2.5 ¢/kWh. Higher conversion efficiencies allow the use of fewer collectors and it is the expense of the latter that dominates the system cost in case I. Only about 4% of the total costs are attributable to the solar cells in this instance and, although this is an extreme case, it does emphasize the fact that cost reductions in the nonsolar cell components of many photovoltaic systems are just as important as reducing cell costs.

One possible problem concerning the operation of photovoltaic central power plants in the above (case (3)) mode is the maintenance of supply during periods of variable cloud cover. For the modest employment of photovoltaic systems that can be expected in the 1985–2000 period coordination with the spinning reserve of nonsolar power plant may be sufficient to preserve the continuity of supply in these instances. However, with the more widespread use of photovoltaic central power plants this method may no longer be feasible and a more satisfactory approach would be to include a small (relative to case (1)) storage facility charged either from the array or from a dedicated nonsolar power source. This three-component arrangement (case (2) above) would be sized such that the rated output power could be continuously provided by the plant, at least for the expected range of insolation conditions. One such system that has recently been analyzed comprises a photovoltaic array, battery store and a coal-fired, backup generation facility rated, in total, at 100 MW and located at Inyokern, California.[296] Results of this analysis are shown in Fig. 6.8, and details of the salient component costs are given in the caption. The values for arrays, storage and power conditioning are lower than used in other examples quoted in this chapter and are intended to be appropriate to the year 1990 and for large systems utilizing advanced technology.[296] The shaded area in Fig. 6.8 refers to the projected price range of conventional fuels as used in Fig. 6.6, and the indications are that a number of array/storage sizes would be cost-competitive with conventional power plant under these circumstances. Lowest cost is predicted for a plant with total array area of approximately 2.5×10^6 m^2 and a storage capacity

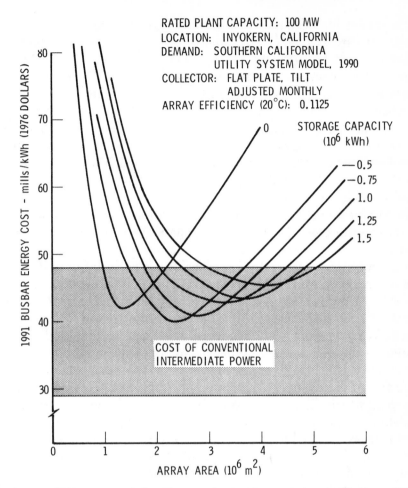

Fig. 6.8. 1990 energy cost (not levelized) of a photovoltaic central power plant
utilizing energy storage and a conventional back-up facility. Costs assumed in the analysis are: array support structure $15 m^{-2}, photovol-
taic arrays $11.25 m^{-2} ($100/kW$_{pk}$), power conditioning $35/kW,
electric storage $20/kWh, coal plant capital costs $550–680/kW. (After Leonard;[296] courtesy of IEEE and National Science Foundation).

of 0.5×10^6 kWh (5 hours use at 100 MW). With this arrangement
the levelized energy cost (not shown in Fig. 6.8) is 4.9 ¢/kWh, the
plant capacity factor is 0.46 and the required coal-fired, backup
capacity is only 25.8 MW.

By way of comparison of the performance of the above plant with
a similarly rated and located plant operating in the no-storage mode

(case (3) at the beginning of this section), it has been estimated that with zero storage and an array area of 1×10^6 m^2 the plant capacity factor would be 0.22, the backup capacity would be 76.1 MW and the levelized energy cost would be again 4.9 ¢/kWh.[296] Although the data given in Ref. 296 do not extend to the case of no backup power capacity (i.e., case (1)), some idea of the performance in this mode can be gained from the results for a storage capacity of 1.5×10^6 kWh and an array area of 5×10^6 m^2, in which case the plant capacity factor was 0.80, the backup capacity only 3.2 MW and the levelized energy cost was 5.8 ¢/kWh. Thus the energy costs resulting from the three modes of operation are all very similar but in the latter case (approximating to base-load operation) cost-competitiveness with conventional power-generating plant, for which base-load costs are much less than intermediate-load costs, is unlikely.[296]

In the photovoltaic power systems discussed so far only the electrical energy output has been considered in determining the systems' worth. However, in systems using very high concentration ratios (~ 1000), as might be possible at the central power station level, there is likely to be some economic advantage to be gained by also utilizing the thermal energy collected on cooling the solar cell arrays. The sale of this energy either in thermal form or as electricity resulting from the driving of an appropriate turbine unit, are two possible ways in which the overall effectiveness of the solar energy conversion might be improved.

Rigorous calculations for hybrid conversion schemes are not yet available owing to present uncertainties in estimating, for example, concentrator costs, high temperature solar cell (GaAs) costs, operation and maintenance costs, and the profitability of selling thermal energy. Preliminary calculations have, however, served to confirm that at $C = 1000$ it is more effective economically to use a system design based on optimizing both electrical and thermal outputs, rather than the photovoltaic output alone.[296] Firm estimates of the cost-effectiveness of hybrid systems should materialize following the commercialization of GaAs solar cells and the transfer of knowledge gained in the operation of high sunlight-concentration schemes for photothermal conversion.

7.

Conclusions

Photovoltaics is already a suitable and cost-effective means of providing electricity for a variety of terrestrial applications. These can be categorized by their small and predictable electrical loads, and often by their remote location or need for independent, reliable and unattended operation. Continued penetration of photovoltaics into markets associated with these applications seems assured, and the experience gained there will assist in reducing the present price of photovoltaic systems' components, particularly of the solar cell arrays for which the technology is presently immature, and in assessing component reliability and lifetime.

A reduction in costs and a knowledge of durability are particularly important if photovoltaic power generation is to progress from its present small usage to incorporation in the larger-scale systems appropriate to the supply of electricity at the residential, commercial and industrial levels. These larger and less predictable loads could, in principle, be met by scaled-up versions of the simple solar cell array-plus-battery storage arrangements which constitute present photovoltaic systems. However, only in high insolation areas and in the cases of energy-conservative dwellings and limited industrial and agricultural processes can the stand-alone operation of photovoltaic power plant be considered. In most other cases such operation is precluded by the large array areas and energy storage capacities that would be required.

If photovoltaic power generation is to be used at all significantly on Earth, then it must be in conjunction with other, nonsolar sources of electricity. In this manner point-of-use operation of photovoltaic plant at the residential, commercial and industrial levels becomes feasible as arrays and storage facilities would not have to be sized for day and night, year-round operation. Whatever space is available could be used for solar collection, conversion and, perhaps, storage, and the remainder of the required power could be supplied from the nonsolar source which, in the vast majority of cases, would be operated by the local utility.

Once the utility enters into the picture the possibility of being able to utilize photovoltaic power on more than just an individual scale emerges; houses, factories, warehouses, parking lots etc. then become possible dispersed generation sites to be coupled into an electrical supply network containing both conventional and photovoltaic central power stations. The benefit to be gained by involving the utility is not just one of diversifying the supply and the demand but also one of more general significance in that the co-ordinated, large-scale implementation of photovoltaics would allow, firstly, the management and prolonged use of existing fossil-fuel reserves and, subsequently, the complementing of nuclear or other base-load generating methods. Photovoltaics becomes progressively more attractive as a large-scale, power generation method as the prices and political and ecological problems of conventional, electricity-generating fuels increase, but photovoltaics deserves more than assumption of widespread use solely by default. Its merits as a bountiful, renewable, nonpolluting and benign energy source should demand that it be considered even if there were no problems associated with "conventional" fuels.

Despite these unquestionable benefits, large-scale implementation of photovoltaics cannot be expected until capital costs are reduced by a factor of about 50. To achieve cost-competitiveness in the residential, commercial and industrial applications discussed in this book it appears that an array price of $100–500/kW$_{pk}$ or about $10–50 m^{-2} is required irrespective of whether the photovoltaic system is dispersed or centrally-located or proprietor- or utility-owned. Under all these circumstances the effective levelized cost at which photovoltaic power could be produced in the United States, for example, is predicted to be in the range of about 4.5–6.5 ¢/kWh depending on the insolation conditions. These costs are of very similar magnitude to the projected busbar energy costs for conventionally-produced electric power in the United States during the period 1985–2000.

A number of technologies are being explored in an attempt to realize these cost goals. In the unconcentrated sunlight approach the present cost-determining factor is that of the solar cells, and its reduction must be commensurate with high volume production and a cell conversion efficiency of about 12–15%. Silicon is still the material most widely expected to be first to satisfy these requirements, even though its absorption properties and bandgap are not optimum for terrestrial photovoltaic energy conversion. The properties of high purity and polished, single crystal wafers, that are required in many silicon electronic devices, may not be needed for solar cells. 10 cm X 10 cm slabs of granular silicon material have already produced 10% efficient solar cells and would presently seem to offer the best prospects for large-area usage, especially in view of the lack of progress in obtaining acceptably efficient cells at high yields from silicon ribbon, i.e., the material form on which many previous forecasts of the success of silicon solar cells were based. Silicon may also have a thin-film solar cell capability in its amorphous form, and this could lead to an economically-attractive product.

In the concentrated sunlight approach to photovoltaic energy conversion it is the solar cell efficiency and the cost of the concentrating, tracking and cooling apparatus that principally determine the array output fixed costs. Silicon is again a strong candidate for the solar cell material and high efficiency (around 15%) operation with cells based on established single crystal wafer technology could be possible at about 100 Suns and, perhaps, 100°C. Operation at higher illumination intensities, but not cell temperatures, is possible with multijunction devices. The other material with excellent credentials for concentrator use is GaAs and efficiencies near the theoretical maximum have been obtained at high illumination levels, i.e., 19% at 1735 Suns. It would seem to be a diminishing return to seek ways of further improving GaAs concentrator cell efficiencies and, although there is room for reduction in cell costs, the important factors in the continued development of GaAs systems are experience and costs associated with mirrors, lenses, cooling and tracking apparatus. The good high temperature performance of GaAs cells may permit operation at about 200°C, so yielding conversion schemes with the potential for significant electrical and thermal energy outputs.

Although the techno-economic goals for the production of cost-effective photovoltaic power appear to be within the limits set by present materials and forseeable technologies, the time of their actual attainment will depend largely on how rapidly and to what extent national governments and international organizations encourage the development of photovoltaics.   Recognition of the benefits and desirability of photovoltaic power generation at the political level would stimulate manufacturers to step up production and utilities and private concerns to increase purchases. These activities would generate cost reductions via volume production and learning experience and enable component reliability to be more firmly assessed. Increased installation of photovoltaic systems would lead to greater public awareness of the practicability of photovoltaics, to more confidence in its use and to further capital investment. Ultimately the commercial attractions of photovoltaics will be large but, more importantly, the benefits to mankind will be significant. The encouragement of photovoltaic power generation *now* will help to ensure the orderly transition from a fossil fuel-based electrical economy to one founded on solar and nuclear sources. In this manner crises in the continuous supply of electrical energy can be averted and succeeding generations throughout the world can be guaranteed the electrical prosperity that so many of us currently enjoy and take for granted.

References

1. Proc. Joint Conf. Am. Sect. ISES and SES Can. Inc., Winnipeg, Aug., 1976.
2. Proc. IEEE Photo. Spec. Conf., 12th, Baton Rouge, La. 1976.
3. DuBow, J. and Curran, L., *Electronics,* pp. 86–99 (Nov. 11 1976).
4. Technical bulletins issued by the following solar cell manufacturers:
   (a) Philips, Lane Cove, N. S. W.
   (b) Solarex, Rockville, Md.
   (c) Sensor Techn. Inc., Chatsworth, Calif.
   (d) Spectrolab, Sylmar, Calif.
   (e) Solar Power Corp., Wakefield, Mass.
5. Moore, W. C., Ravin, J. W., Masters, R. M., and Forestieri, A. F., Proc. IEEE Photo. Spec. Conf., 10th, Palo Alto, Calif. p. 227, 1973.
6. Rosenblatt, A. I., *Electronics,* pp. 99–111 (April 4 1974).
7. Faehn, D. D., in Ref. 2, p. 715.
8. *Electrical World,* p. 42 (Sept. 1, 1976).
9. Merrigan, J. A., "Sunlight to Electricity," MIT Press: Cambridge, Mass., 1975.
10. Polgar, S., Proc. UNESCO Inter. Congr., "The photovoltaic power and its applications in space and on Earth," Paris, p. 554, 1973.
11. Magid, L. M., in Ref. 2, p. 607.
12. Treble, F. C., private communication, Nov., 1976.
13. Lindmayer, J., quoted in Ref. 3, p. 92.
14. Bond, J. W. Jr., in Ref. 2, p. 725.
15. Daniels, F., "Direct Use of The Sun's Energy," Yale University Press: New Haven, Mass, 1964.
16. Usmani, I. H., presentation at IEEE Photo. Spec. Conf., 12th, Baton Rouge, La., Nov. 1976, (unpublished).
17. Williams, J. R., *Astron. and Aeronaut.,* p. 46 (Nov. 1975).
18. Boes, E. C., Hall, I. J., Prairie, R. R., Stromberg, R. P., and Anderson, H. E., in Ref. 1, 1, p. 238.
19. Bartels, F. T. C. and Moffett, D. W., Proc. IEEE Photo. Spec. Conf., 10th, Palo Alto, Calif. p. 258, 1973.
20. Boes, E. C., Sandia Labs. Rept. No. SAND 76–0009, Feb. 1976.
21. Backus, C. E., *J. Vac. Sci. Technol.* 12, 1032, 1975.
22. Böer, K. W., in Ref. 1, 1, p. 264.
23. Thekaekara, M. P., *Appl. Optics* 13, 518, 1974.
24. Moon, P., *J. Franklin Inst.,* 230, 583, 1940.
25. Intern. Commission on Illumination, CIE No. 20 (TC-2.2), Bureau Central de la CIE, Paris, 1972.

26. Thomas, A. P. and Thekaekara, M. P., in ref. 1, 1, p. 338.
26a. Brandhorst, H. W., Jr. Presentation at IEEE Photo. Spec. Conf., 12th, Baton Rouge, La. Nov., 1976, (unpublished).
27. Burgess, E. L. and Edenburn, M. W., in Ref. 2, p. 774.
28. Cherdak, A. S. and Haas G. M., in Ref. 2., p. 794.
29. Kirpich, A. and Buerger, E., in Ref. 2, p. 673.
30. Shepard, N. F. Jr. and Landes, R., in Ref. 2, p. 705.
31. Backus, C. E., Proc. Intersoc. Energy Conv. Eng. Conf., 7th, p. 704, 1972.
32. Yasui, R. K. and Patterson, R. E., Proc. IEEE Photo. Spec. Conf., 10th, Palo Alto, Calif. p. 238, 1973.
33. Carmichael, D. C., Gaines, G. B., Sliemers, F. A. and Kistler, C. W., in Ref. 2, p. 317.
34. Buhs, R. and Gochermann, H., in Ref. 2, p. 634.
35. Carroll, W. F. and Cuddihy, E., in Ref. 2, p. 332.
36. Kirkpatrick, A. R., Kreisman, W. S., and Minnucci, J. A., in Ref. 2, p. 573.
37. Ross, R. G. Jr., in Ref. 2, p. 801.
38. Pryor, R. A., Coleman, M. G. and Keeling, M. C., in Ref. 2, p. 375.
39. Roger, J., Proc. IEEE Photo. Spec. Conf., 10th, Palo Alto, Calif., p. 344, 1973.
40. Ross, R. G., Jr., in Ref. 1, 6, p. 48.
41. Cherry, W. R., *Trans. Am. Soc. Mech. Eng.* A94, 78, 1972.
42. Gorski, A., Graven, R., McIntire, W., Schertz, W. W., Winston, R., and Zwerdling, S., in Ref. 2, p. 764.
43. Backus, C. E., Proc. Intersoc. Energy Conv. Eng. Conf., 10th, p. 601, 1975.
44. Mash, D. H., quoted in *Sunworld,* 1 (1), p. 22, July 1976.
45. Bartels, F. T. C. and Kelber, C. C., in Ref. 2, p. 691.
46. Rabl, A., *Sol. Energy,* 18, 93, 1976.
47. Goodman, N. B., Ignatius, R., Wharton, L., and Winston, R., in Ref. 1, 6, p. 325.
48. Backus, C. E., Final report to NSF, Grant No CI-41894, Jan. 1975.
49. Florschuetz, L. W., in Ref. 104, p. 318.
50. Backus, C. E., Semi-annual progress rept. to Sandia Labs., Contract No. 02-7850, July 1976.
51. James, L. W., Moon, R. L., Moore, E. O., and Bell, R. L., in Ref. 2, p. 771.
52. Mlavsky, A. I., Serreze, H. B., Stormont, R. W., and Taylor, A. S., in Ref. 2, p. 160.
53. Haas, G. M. and Bloom, S., in Ref. 104, p. 256.
54. Federmann, E. F., Ferber, R. R., Pittman, P. F., and Chowaniec, C. R., Proc. Intersoc. Energy Conv. Eng. Conf., 11th, p. 1308, 1976.
55. "Energy Storage," Exec. Summary of Engineering Foundation Conf., Asilomar, Feb. 1976. Available as CONF-760212 from NTIS, U. S. Dept. of Commerce, Va. 22161.
56. Braun, C., Cherniavsky, E. A., and Salzano, F. J., Proc. Intersoc. Energy Conv. Eng. Conf., 10th, p. 82, 1975.
57. Bockris, J. O'M., "Energy: The Solar-Hydrogen Alternative," Architectural Press: London, 1976.
58. Douglas, D. L., IEEE Power Eng. Soc. papers (74 CH0913-4-PWR), p. 37, 1974.
59. Yao, N. P., Presentation at Jt. Conf. Am. Sec. ISES and SES Can. Inc., Winnipeg, Aug., 1976, (unpublished).

60. Rogers, F. C. and Allen, A. E., IEEE Power Eng. Soc. papers (74 CH0913-4-PWR), p. 21, 1974.
61. Kalhammer, F. R., in Ref. 55, p. 37.
62. Ramakumar, R., in Ref. 1, 8, p. 163.
63. Lueckel, W. J. and Farris, P. J., IEEE Power Eng. Soc. papers (74 CH0913-4-PWR), p. 42, 1974.
64. King, J. M. Jr., Proc. Intersoc. Energy Conv. Eng. Conf., 10th, p. 237, 1975.
65. Fullman, R. L., Proc. Intersoc. Energy Conv. Eng. Conf., 10th p. 91, 1975.
66. Notti, J. E., Jr., in Ref. 55, p. 30.
67. Segal, H. R. and Boom, R. W., Proc. Intersoc. Energy Conv. Eng. Conf., 10th p. 101, 1975.
68. Hassenzahl, W. V., in Ref. 55, p. 32.
69. Kelly, B. P., Eckert, J. A., and Berman, E., Proc. UNESCO Inter. Cong., "The photovoltaic power and its applications in space and on Earth," Paris, p. 551 (1973).
70. Glaser, P. E., *Science* 162, 857, 1968.
71. Brown, W. C. and Maynard, O. E., Proc. AIAA/AAS Solar Energy for Earth Conf., Paper No. 75-642, 1975.
72. Glaser, P. E., in Ref. 1, 1, p. 1.
73. Oman, H., in Ref. 2, p. 832.
74. Kline, R. and Nathan, C. A., Proc. AIAA/AAS Solar Energy for Earth Conf., Paper No. 75-641, 1975.
75. Ehricke, K. A., Rockwell Intern. Corp. Rept. No. E74-3-1, Mar. 1974.
76. Ramakumar, R., in Ref. 1, 9, p. 162.
77. Clifford, A., in Ref. 2, p. 826.
78. Thomas, W. A., presentation at Am. Assoc. Adv. Sci. meeting, Boston, Mass., Feb. 1976.
79. Robbins, R. L., *Sol. Energy* 18, 371, 1976.
80. Bezdek, R. H. and Maycock, P. D., in Ref. 1, 9, p. 64.
81. Lorsch, H. G., in Ref. 1, 9, p. 97.
82. Becquerel, E., Compt. Rend. 9, 561, 1839.
83. Adams, W. G. and Day, R. E., *Proc. Roy. Soc.* A25, 113, 1877.
84. Lange, B., *Zeit. Phys.* 31, 139, 1930.
85. Grondhal, O., *Rev. Mod. Phys.* 5, 141, 1933.
86. Schottky, W., *Zeit. Phys.* 31, 913, 1930.
87. Chapin, D. M., Fuller, C. S. and Pearson, G. L., *J. Appl. Phys.* 25, 676, 1954.
88. Kirkpatrick, A. R., in Ref. 1, 6, p. 67.
89. Christensen, O. and Bay, H. L., *Appl. Phys. Lett.* 28, 491, 1976.
90. Wagner, S., *J. Crystal Growth* 31, 113, 1975.
91. Vohl, P., Perkins, D. M., Ellis, S. G., Addiss, R. R., Hui, W., and Noel, G., *IEEE Trans. Elec. Dev.* ED-14, 26, 1967.
92. Wolf, M., *J. Vac. Sci. Technol.* 12, 984, 1975.
93. Wolf, M. and Rauschenbach, H., *Advan. Energy Conv.* 3, 455, 1963.
94. Lindholm, F. A., Fossum, J. G., and Burgess, E. L., in Ref. 2, p. 33.
95. Prince, M. B., *J. Appl. Phys.* 26, 534 1955.
96. Lindholm, F. A. and Neugroschel, A., in Ref. 1, 6, p. 120.
97. Fossum, J. G., *Sol.-State Electron.* 19, 269, 1976.
98. Dunbar, P. M. and Hauser, J. R., *Sol.-State Electron.* 19, 95, 1976.

99. Sutherland, J. E. and Hauser, J. R., in Ref. 2, p. 939.
100. Tsaur, S. C. Milnes, A. G., Sahai, R., and Feucht, D. L., Proc Int. Symp on GaAs, Inst. of Physics, London, p. 156, 1972.
101. McOuat, R. F. and Pulfrey, D. L., *J. Appl. Phys.* 47, 2113, 1976.
102. Lanza, C. and Hovel, H. J., in Ref. 2, p. 96.
103. Hovel, H. J., "Solar Cells," vol. 11 of Semiconductors and Semimetals, Beer, A. C. and Willardson, R. K., eds., Academic Press: N.Y., 1975.
104. Proc. IEEE Photo. Spec. Conf., 11th, Scottsdale, Ariz., 1975.
105. Iles, P. A. and Soclof, S. I., in Ref. 104, p. 19.
106. Johnston, W. D., Jr. and Callahan, W. M., *Appl. Phys. Lett.* 28, 150, 1976.
107. Milnes, A. G. and Feucht, D. L., "Heterojunctions and Metal-Semiconductor Junctions," Academic Press: N. Y. 1972.
108. See Ref. 103, p. 130.
109. Hutchby, J. A. and Fudurich, R. L., *J. Appl. Phys.* 47, 3140, 1976.
110. Hutchby, J. A. and Fudurich, R. L., *J. Appl. Phys.* 47, 3152, 1976.
111. Böer, K. W., in Ref. 1, 6, p. 130.
112. Böer, K. W., *Phys. Rev.* B13, 5373, 1976.
113. Barnett, A. and Rothwarf, A. in Ref. 2, p. 544.
114. Fowler, R. H., *Phys. Rev.* 38, 45, 1931.
115. McOuat, R. F., and Pulfrey, D. L., in Ref. 104, p. 371.
116. Card, H. C. and Yang, E. S., *Appl. Phys. Lett.* 29, 51 1976.
117. Viktorovitch, P., Kamarinos, G. and Even, P., in Ref. 2, p. 870.
118. Hovel, H. J., *J. Appl. Phys.* 47, 4968, 1976.
119. Lindmayer, J. and Allison, J. F., *Comsat Tech. Rev.* 3, 1, 1973.
120. Baroana, C. R. and Brandhorst, H. W., Jr., in Ref. 104, p. 44.
121. Restrepo, F. and Backus, C. E., *IEEE Trans. Elec. Dev.* ED-23, 1195, 1976.
122. Arndt, R. A., Allison, J. F., Haynos, J. G., and Meulenberg, A., Jr., in Ref. 104, p. 40.
123. Franz, S., Kent, G., and Anderson, R. L., *J. Electronic Mater.* 6, 107, 1977.
124. DuBow, J. B., Burk, D. E., and Sites, J. R., *Appl. Phys. Lett.* 29, 494, 1976.
125. Fabre, E., Michel, J., and Baudet, Y., in Ref. 2, p. 904.
126. Kamath, G. S., Ewan, J., and Knechtli, R. C., in Ref. 2, p. 929.
127. Konagai, M. and Takahashi, K., *J. Appl. Phys.* 46, 3542, 1975.
128. Johnston, W. D., Jr., and Callahan, W. M., in Ref. 2, p. 934.
129. Wang. E. Y. and Legge, R. N., in Ref. 2, p. 967.
130. Stirn, R. J. and Yeh, Y. C. M., *Appl. Phys. Lett.* 27, 95, 1975.
131. Yamaguchi, K., Matsumoto, H., Nakayama, N., and Ikegami, S., *Jap. J. Appl. Phys.* 15, 1575, 1976.
132. Shay, J. L., Wagner, S., Bachmann, K. J., Bueler, E., and Kasper, H. M., in Ref. 104, p. 503.
133. Shay, J. L., Wagner, S., Bachmann, K. J., and Bueler, E., *J. Appl. Phys.* 47, 614, 1976.
134. Meakin, J. D., Baron, B., Böer, K. W., Burton, L. C., Devaney, W., Hadley, H., Jr., Phillips, J., Rothwarf, A., Storti, G., and Tseng, W., in Ref. 1, 6, p. 113.

135. Burton, L. C., Baron, B., Devaney, W., Hench, T., Lorenz, S., and Meakin, J. D., in Ref. 2, p. 526.
136. Luquet, H., Szepessy, L., Bougnot, J., Savelli, M., and Guastavino, F., in Ref. 104, p. 445.
137. Green, M. A., King, F. D. and Shewchun, J., *Sol.-State Electron.* 17, 551, 1974.
138. Sze, S. M., "Physics of Semiconductor Devices," J. Wiley: N. Y., 1969.
139. Godlewski, M. P., Brandhorst, H. W., Jr., and Baroana, C. R., in Ref. 104 p. 32.
140. Lindholm, F. A., Neugroschel, A., Sah, C. T., Godlewski, M. P., and Brandhorst, H. W., Jr., in Ref. 2, p. 1.
141. Mandelkorn, J. and Lamneck, J. H., in Ref. 104, p. 36.
142. Hovel, H. J., Proc. IEEE Photo. Spec. Conf. 10th, Palo Alto, Calif., p. 34, 1973.
143. See Ref. 103, p. 54.
144. Fahrenbruch, A. L. and Bube, R. H., *J. Appl. Phys.* 45, 1264, 1974.
145. Rhoderick, E. H., *J. Phys. D.* 3, 1153, 1970.
146. Pulfrey, D. L., *IEEE Trans. Elec. Dev.* ED-23, 587, 1976.
147. Fonash, S. J., *J. Appl. Phys.* 46, 1286, 1975 and 47, 3597, 1976; and in Ref. 104, p. 376.
148. Green, M. A. and Godfrey, R. B., *Appl. Phys. Lett.* 29, 610, 1976.
149. Parrott, J. E., *IEEE Trans. Elec. Dev.* ED-21, 89, 1974.
150. Fischer, H. and Pschunder, W., in Ref. 104, p. 25.
151. D'Aiello, R. V., Robinson, P. H., and Kressel, H., *Appl. Phys. Lett.* 28, 231, 1976.
152. Hovel, H. J. and Woodall, J. M., *J. Electrochem. Soc.* 120, 1246, 1973.
153. Hovel, H. J. and Woodall, J. M., in Ref. 2, p. 945.
154. Alferov, Zh. I., Andreev, V. M., Kagan, M. B., Protasov, I. I., and Trofim, V. G., *Sov. Phys. Semicon.* 4, 2047, 1971.
155. Stirn, R. J. and Yeh, Y. C. M., in Ref. 2, p. 883.
156. Shay, J. L., Bettini, M., Wagner, S., Bachmann, K. J., and Buehler, E., in Ref. 2, p. 540.
157. Fahrenbruch, A. L., Buch, F., Mitchell, K. W., and Bube, R. H., in Ref. 2 p. 529.
158. Stirn, R. J., Proc. IEEE Photo. Spec. Conf., 9th, Silver Spring, Md., p. 72, 1972.
159. Lindmayer, J., *Comsat. Techn. Rev.* 2, 105, 1972.
160. Wolf, M., Proc. IEEE Photo. Spec. Conf., 10th, Palo Alto, Calif., p. 5, 1973.
161. Anderson, R. L., *Appl. Phys. Lett.* 27, 691, 1975.
162. Pulfrey, D. L., *Sol.-State Electron.* 20, 455, 1977.
163. Stirn, J. R. and Yeh, Y. C. M., in Ref. 104, p. 391.
164. Pulfrey, D. L. and McOuat, R. F., *Appl. Phys. Lett.* 24, 167, 1974.
165. Fahrenbruch, A. L., Vasilchenko, V., Buch, F., Mitchell, K., and Bube, R. H., *Appl. Phys. Lett.* 25, 605, 1974.
166. Sreedhar, A. K., Sharma, B. L., and Purohit, R. K., *IEEE Trans. Elec. Dev.* ED-16, 309, 1969.
167. Green, M. A., *IEEE Trans. Elec. Dev.* ED-23, 11, 1976.
168. Dunbar, P. M. and Hauser, J. R., in Ref. 2, p. 23.

169. Singh, R. and Shewchun, J., *Appl. Phys. Lett.* **28**, 512, 1976.
170. James L. W. and Moon, R. L., in Ref. 104, p. 402.
171. Hovel, H. J., in Ref. 1, 6, p. 1.
172. Demeo, E. A., EPRI Rept. No ER-188, Feb. 1976.
173. Field, M. B. and Scudder, L. R., in Ref. 2, p. 303.
174. Haigh, A. D., in Ref. 2, p. 360.
175. Matzen, W. T., Chiang, S. Y., and Carbajal, B. G., in Ref. 2, p. 340.
176. Iles, P. A. and Soclof, S. I., in Ref. 2, p. 978.
177. Boone, J. L. and van Doren, T. P., in Ref. 1, **10**, p. 212.
178. See Ref. 103, p. 64.
179. Shewchun, J., Green, M. A. and King, F. D., *Sol.-State Electron* **17**, 563, 1974.
180. Green, M. A., private communication, Feb. 1976.
181. Wolf, H. F., "Semiconductors," J. Wiley: N. Y., 1971.
182. Aukerman, L. W., Millea, M. F., and McColl, M., *J. Appl. Phys.* **38**, 685, 1967.
183. Redfield, D., in Ref. 104, p. 431.
184. Jordan, J. F., in Ref. 104, p. 508.
185. Rothwarf, A., in Ref. 2, p. 488.
186. Soclof, S. I. and Iles, P. A., in Ref. 104, p. 56.
187. Fischer, H. and Pschunder, W., in Ref. 2, p. 86.
188. Lindmayer, J., in Ref. 2, p. 82.
189. Cuomo, J., DiStefano, T. H., and Rosenberg, R., IBM Techn. Discl. Bull **17**, 2455, 1975.
190. Iles, P. A., quoted in Ref. 103, p. 108.
191. Hunt, L. P., in Ref. 2, p. 347.
192. Wakefield, G. F., Maycock, P. D., and Chu, T. L., in Ref. 104, p. 49.
193. Hill, D. E., Gutsche, H. W., Wang, M. S., Gupta, K. P., Tucker, W. F., Dowdy, J. D., and Crepin, R. J., in Ref. 2, p. 112.
194. Davis, J. R., Rai-Choudhury, P., Blais, P. D., Hopkins, R. H., and McCormick, J. R., in Ref. 2, p. 106.
195. Browning, M. F., Blocher, J. M., Wilson, W. J., and Carmichael, D. C., in Ref. 2, p. 130.
196. Hunt, L. P., Dosaj, V. D., McCormick, J. R., and Crossman, L. D., in Ref. 2, p. 125.
197. McCormick, J. R., Crossman, L. D., and Raucholz, A., in Ref. 104. p. 270.
198. Rea, S. N. and Wakefield, G. F., in Ref. 1, 6, p. 57.
199. Schmid, F., in Ref. 2, p. 146.
200. Schmid, F., *Electronics*, p. 34 (June 24 1976).
201. Zoutendyk, J. A., in Ref. 1, 6, p. 34, and *Solar Energy* (in press).
202. Seidensticker, R. G., Scudder, L., and Brandhorst, H. W., Jr., in Ref. 104, p. 299.
203. Ravi, K. V. and Mlavsky, A. I., in Ref. 1, 6, p. 23: and references quoted therein.
204. Bates, H. E., Jewett, D. N., and White, V. E., Proc. IEEE Photo. Spec. Conf., 10th, Palo Alto, Calif., p. 197, 1973.
205. Kressel, H., D'Aiello, R. V., and Robinson, P. H., *Appl. Phys. Lett.* **28**, 157, 1976.
206. Hari Rao, C. V., Bates, H. E. and Ravi, K. V., *J. Appl. Phys.* **47**, 2614, 1976.

207. Digges T. G. Jr., Leipold M. H., Koliwad K. M., Turner G. and Cumming G. D., in Ref. 2, p. 120.
208. Lesk, I. A., Baghdadi, A., Gurtler, R. W., Ellis, R. J., Wise, J. A., and Coleman, M. G., in Ref. 2, p. 173.
209. Fang, P. H., Progress Rept. No. NSF/RANN/SE/GI-34975/73/4, NSF Grant No. GI-34975, Jan., 1974.
210. Fang, P. H., Ephrath, L., and Nowak, W. B., *App. Phys. Lett.* 25, 583, 1974.
211. Chu, T. L., Mollenkopf, H. C., and Chu, S. S., *J. Electrochem. Soc.* 122, 1681, 1975.
212. Chu, T. L., *J. Vac. Sci. Technol.* 12, 912, 1975.
213. Chu, T. L., Chu, S. S., Duh, K. Y., and Yoo, H. I., in Ref. 2, p. 74.
214. Milnes, A. G. and Feucht, D. L., in Ref. 104, p. 338.
215. Konagai, M. and Takahashi, K., Ext. Abstr. No. 224, Electrochem. Soc. Spring meeting, Wash., D.C., 1976.
216. Minagawa, S., Saitoh, T., Warabisako, T., Nakamura, N., Itoh, H., Tamura H., and Tokuyama, T., in Ref. 2, p. 77.
217. Milnes, A. G. and Feucht, D. L., in Ref. 2, p. 997.
218. Heaps, J. D., Maciolek R. B., Zook J. D., and Scott M. W., in Ref. 2, p. 147.
219. Griffith, R. W., in Ref. 1, 6, p. 205.
220. Carlson, D. E. and Wronski, C. R., *Appl. Phys. Lett.* 28, 671, 1976.
221. Spear, W. E., le Comber, P. G., Kinmond, S. and Brodsky, M. H., *Appl. Phys. Lett.* 28, 105, 1976.
222. Carlson, D. E., Wronski, C. R., Triano, A. R., and Daniel, R. E., in Ref. 2, p. 893.
223. Ralph, E., in Ref. 104, p. 315.
224. Coleman, M. G., Bailey, W. L., and Pryor, R. A., in Ref. 2, p. 313.
225. Shah, P. and Fuller, C. R., in Ref. 2, p. 286.
226. Chandler, T. C., Hilbron, R. B., and Faust, J. W., in Ref. 2, p. 282.
227. Wichner, R. and Charlson, E. J., *J. Electron. Mater.* 5, 513, 1976.
228. Chen, L. Y. and Loferski, J. J., Final contract rept., NASA Grant No. NGR-40-002-982, June 1975.
229. Salter, G. C. and Thomas, R. E., in Ref. 104, p. 364.
230. Norman, C. E. and Thomas, R. E., in Ref. 2, p. 993.
231. van Halen, P., Thomas, R. E., Mertens, R., and van Overstraeten, R., in Ref. 2, p. 907.
232. Card, H. C., *IEEE Trans. Elec. Dev.* ED-23, 538, 1976.
233. Nash, T. R. and Anderson, R. L., in Ref. 2, p. 975; and *IEEE Trans. Elec. Dev.* ED-24, 468, 1977.
234. Franz, S. L., Thompson, W. G., Kent, G., and Anderson, R. L., Proc. NSF/ERDA National Workshop on Low Cost Polycrystalline Si Solar Cells, Dallas, May 1976.
235. Reynolds, D. C., Leies, G., Antes, L. L., and Masbruger, R. E., *Phys. Rev.* 96, 533, 1954.
236. TeVelde, T. S. and Dieleman, J., *Phil. Res. Repts.* 28, 573, 1973.
237. Wyeth, N. C. and Catalano, A. W., in Ref. 2, p. 471.
238. Besson, J., Nguyen Duy, T., Gauthier, A., Palz, W., Martin, C., and Vedel, J., in Ref. 104, p. 468.
239. Pfisterer, F., Schock, H. W., and Bloss, W. H., in Ref. 2, p. 502.
240. Hsieh, E. J., Proc. Int. CdS Solar Cell Workshop, 1st, Newark, 1975.

241. Shewchun, J., Loferski, J. J., Wold, A., Arnott, R., DeMeo, E. A., Beulieu, R., Wu, C. C., and Hwang, H. L., in Ref. 104, p. 482.
242. Arjona, F., Rueda, F., Garcia-Camarero, E., León, M. and Arizmendi, L., in Ref. 2, p. 515.
243. Bougnot, J., Perotin, M., Marucchi, J., Sirkis, M., and Savelli, M., in Ref. 2, p. 519.
244. Biter, W. J. and Shirland, F. A., in Ref. 2, p. 466.
245. DiZio, S. F., in Ref. 1, 6, p. 108.
246. Nakayama, N., Matsumoto, H., Yamaguchi, K., Ikegami, S., and Hioki, Y., *Jap. J. Appl. Phys.* 15, 2281, 1976.
247. Kazmerski, L. L., White, F. R., Sanborn, G. A., Merrill, A. J., Ayyagari, M. S., Mittleman, S. D., and Morgan, G. K., in Ref. 2, p. 534.
248. Olsen, L. C. and Bohara, R., in Ref. 104, p. 381.
249. Trivich, D., Wang. E. Y., Komp, R. J., and Ho, F., in Ref. 2, p. 875.
250. Warschauer, D., in Ref. 2, p. 613.
251. Treble, F. C., in Ref. 2, p. 625.
252. Rodot, M. and Palz, W., in Ref. 2, p. 618.
252a. Bachman, K. J., Bueler, E., Shay, J. L., and Wagner, S., *Zeit. Phys. Chem. N. F.*, 98, 365, 1975.
253. Cusano, D. A., *Sol.-State Electron.* 6, 217, 1963.
254. Bell, R. O., Serreze, H. B., and Wald, F. V., in Ref. 104, p. 497.
255. Russel, B. G. and Pulfrey, D. L., in Ref. 2, p. 962.
256. See Ref. 103, p. 217.
257. Assimos, J. A. and Trivich, D., *J. Appl. Phys.* 44, 1687, 1973.
258. Toth, R. S., Kilkson, R., and Trivich, D., *J. Appl. Phys.* 31, 1117, 1960.
259. Drobny, V. F. and Pulfrey, D. L., unpublished data.
260. Forman, R., research proposal submitted to NSF, Apr. 1975, (unpublished.)
261. Alvi, N. S., Backus, C. E., and Masden, G. W., in Ref. 2, p. 948.
262. Loferski, J. J., in Ref. 2, p. 957.
263. See Ref. 103, pp. 211–16.
264. Reucroft, P. J., Takahashi, K., and Ullal, H., *Appl. Phys. Lett.* 25, 664, 1974.
265. Reucroft, P. J., Takahashi, K., and Ullal, H., *J. Appl. Phys.* 46, 5218, 1975.
266. Merritt, V. Y. and Hovel, H. J., *Appl. Phys. Lett.* 29, 414, 1976.
267. Fang. P. H., *J. Appl. Phys.* 45, 4672, 1974.
268. Lindmayer, J., in Ref. 104, p. 317.
269. Fossum, J. G. and Burgess, E. L., in Ref. 2, p. 737.
270. Backus, C. E., editor, "Solar Cells," p. 9, IEEE Press: N. Y., 1976.
271. Castle, J. A., in Ref. 2, p. 751.
272. Gray, P. E., *IEEE Trans. Elec. Dev.* ED-16, 424, 1969.
273. Dhariwal, S. R., Kothari, L. S., and Jain S. C., *IEEE Trans. Elec. Dev.* ED-23, 504, 1976.
274. Vernon, S. M. and Anderson, W. A., *Appl. Phys. Lett.* 26, 707, 1975.
275. Thompson, W. G., Franz, S. L., Anderson, R. L., and Winn, O. H., *IEEE Trans. Elec. Dev.* ED-24, 463, 1977.
276. Ewan, J., Kamath, G. S., and Knechtli, R. C., in Ref. 104, p. 409.
277. Shen, C. C. and Pearson, G. L., Proc. IEEE Int. Electron Dev. meeting, p. 99, Wash., D.C., 1975.

278. Ettenberg, M. and Nuese, C. J., *J. Appl. Phys.* **46**, 3500, 1975.
279. Bell, R. L., quoted in *Electronics*, p. 42, (May 29, 1975).
280. Dean, R. H., Napoli, L. S., and Liu, S. G., *RCA Rev.* **36**, 324, 1975.
281. Rule, T. T., Harmon, S. Y., Backus, C. E., and Jacobson, D. L., in Ref. 2, p. 744.
282. Fossum, J. G. and Lindholm, F. A., *IEEE Trans. Elec. Dev.* **ED-24**, 325, 1977.
283. See Ref. 103, pp. 139–144, and references quoted therein.
284. Shah, P., *Sol.-State Electron.* **18**, 1099, 1975.
285. Lloyd, W. W., in Ref. 104, p. 349.
286. Sater, B. L. and Goradia, C., in Ref. 104, p. 356.
287. Goradia, C., Ziegman, R., and Sater, B. L., in Ref. 2, p. 781.
288. Soukup, R. J., *J. Appl. Phys.* **48**, 440, 1977.
289. Schwartz, R. J. and Lammert, M. D., Proc. IEEE Int. Electron Dev. meeting, p. 350, Wash., D.C., 1975.
290. Matsushita, T. and Mamine, T., Proc. IEEE Int. Electron Dev. meeting, p. 353, Wash., D.C., 1975.
291. Böer, K. W., in Ref. 1, 9, p. 1.
292. Moore, R. M., *Solar Energy* **18**, 225, 1976.
293. Evans, D. L. and Florschuetz, L. W., *Solar Energy* **19**, 255, 1977.
294. Doane, J. W., O'Toole, R. P., Chamberlain, R. G., Bos, P. B., and Maycock, P. D., "The cost of energy from utility-owned solar electric systems," ERDA/JPL Report No. 1012-76/3, June, 1976.
295. Evans, D. L. and Florschuetz, L. W., paper submitted to *Solar Energy*.
296. Leonard, S. L. in Ref. 2, p. 641.
297. Kirpich, A., Shepard, N. F. Jr., and Irwin, S. E., Proc. Intersoc. Energy Conv. Eng. Conf., 11th, p. 1300, 1976.
298. Schueler, D. G. and Marshall, B. W., in Ref. 2, p. 661.
299. Costogue, E. N. and Lindena, S., Proc. Intersoc. Energy Conv. Eng. Conf., 11th, p. 1449, 1976.
300. Yeh, Y. C. M., Ernest, F. P. and Stirn, R. J., *J. Appl. Phys.* **47**, 4107, 1976.
301. Sater, B. L. and Goradia, C., Proc. Intersoc. Energy Conv. Eng. Conf., 11th, p. 1316, 1976.
302. Yoshikawa A. and Sakai Y., *Sol.-State Electron.* **20**, 133, 1977.
303. Shay, J. L., Wagner S., Bettini M., Bachmann K. J., and Buehler E., *IEEE Trans. Elec. Dev.* **ED-24**, 483, 1977.
304. Koyanagi T., in Ref. 2, p. 627.
305. Kran A., Proc. Intersoc. Energy Conv. Eng. Conf., 11th, p. 1324, 1976.
306. Wronski, C. R., *IEEE Trans. Elec. Dev.* **ED-24**, 351, 1977.
307. Sah C. T. and Lindholm F. A., *IEEE Trans. Elec. Dev.* **ED-24**, 358, 1977.
308. Milnes A. G., Feucht D. L., and Ouyang G., Proc. NSF/ERDA *National Workshop on Low Cost Polycrystalline Silicon Solar Cells, Dallas*, p. 430, 1976.

Index

Index

Absorption, photon, 67–68, 72–91, 110
in $Ga_{1-x}Al_xAs$, 161, 163
in GaAs, 68, 120
in organics, 152
in Si, 68, 137, 166
in thin films, 115, 120, 147, 152, 166
Abundance, solar cell materials
As, 150, 163
Cd, 143
Ga, 150, 163
Si, 115
Air mass, 18
effect on limit conversion efficiency, 109
Amorphous semiconductors, 115
Si, 136–137, 199
Antireflection coating, 71, 137
for GaAs cells, 163–164
for Schottky barrier cells, 90–92, 104,
113, 150
for Si cells, 91, 110, 167, 169
Aphodid burner, 45, 49
Arrays, 2, 9–10, 21–38, 65, 174
concentrator, 28–37, 163, 175–179, 199
cost, 114, 175–179, 184–186, 198
efficiency loss due to mismatching, 39–40
flat plate, 21–27, 52, 114, 177
of CdS cells, 143–144, 177–178
of GaAs cells, 163, 179, 199
of Si cells, 133, 176–179
orientation, 10, 16, 21, 28, 52, 177
production, 124
sizing, 52, 54–56, 180, 196–197
space, 59–60
tracking, 4, 16, 28, 31–33, 35, 155, 189
typical characteristics, 53

Back surface field, 97–98
in CdTe cells, 146

in GaAs cells, 112, 158, 162
in Schottky barrier cells, 101, 103, 141
in Si cells, 119–120, 160, 165, 167
optimization, 110
Bandgap shrinkage, 97
Base resistivity
effect on efficiency, 97
interface properties, 80
J_{sc}, 77, 79, 87–89
V_{oc}, 95–98, 103, 110, 113
in Si concentrator cells, 160, 165–167,
169
Batteries, storage, 9, 37–38, 42–47, 51, 184
charge control, 39
cost, 194
sizing, 52, 54, 56, 195
See also Energy storage
Blocking diode, 9, 38, 51–52

Cadmium sulfide cells, 116, 141–147
absorption in, 69, 84, 121
arrays, 143–144, 177–178
dark currents, 93, 99–100
efficiency, 111, 113, 141
fabrication, 141–147
fill factor, 108
J_{sc}, 84–85, 92
stability, 141, 143–147
V_{oc}, 104
Zn doping of, 92, 100, 104, 111, 113,
141–142, 144
Cadmium telluride, 120, 145–146, 148
CdTe/CdS cells, 92, 104, 108, 112–113,
116, 142
Chemical vapor deposition, 138
Collection efficiency, 81, 83
Compressed air storage, 45, 48

Concentration, sunlight, 4, 7, 16, 199
  array costs, 28, 37, 174–179
  arrays suitable for, 28–37, 57, 163, 165
  cells suitable for, 32–33, 154–170, 199
    GaAs, 33, 162–164, 196, 199
    Si, multijunction, 167–170
    Si, single junction, 32–33, 164–166
  effect on efficiency, 156–162
  ratio, 28, 30–33, 154–170
Conductivity modulation, 157–158, 169
Contact grid, 71, 107, 116–117, 137, 165
  for CdS cells, 108, 143
  for GaAs cells, 163
  for Si cells, 91, 156, 164–165, 169
Conventional fuels, 1, 186, 194, 198
  costs of, 186, 194
Cooling of cells/arrays, 4, 33, 36, 155, 176
Cost, $/W of cell/array, 3–5, 114, 153–154,
  171–198
  assembly, 23
  CdS, 143–144
  concentrators, 28, 37, 154, 174–179
  GaAs, 163
  installation, 26, 171, 174–179
  Si, 26, 128, 138
  U.S. goals, 3, 5
Cost, photovoltaic electricity, 3, 63,
  171–196, 198
Cu_2O, 66, 116, 148–150
$CuInS_2$, 142, 145–147
$CuInSe_2$, 92, 112, 116, 142, 145–147
Czochralski, 126, 128–129

Dark current, 70, 72, 93, 110
  effect of sunlight concentration, 155, 157,
  159
  in heterojunctions, 98–100, 104, 112
  in homojunctions, 94–98
  in low cost cells, 118
  in polycrystalline material, 122
  in Schottky barrier cells, 100–104, 110
  119
  relationship to R_{sh}, 117
Dendritic web growth of Si, 129–130
Diffusion potential, 103–104, 111, 113, 157
Diode exponential factor, 93, 100, 104–107
Dip coating semiconductor growth
  of Cu_2S, 142–143
  of Cu_2Te, 146
  of Si, 135–136
Dispatch of photovoltaic electricity, 11, 43,
  189
Displacement
  energy factor, 21, 38, 184, 187
  of conventional fuel, 64, 172, 193
  of conventional generating capacity, 4, 11,
  64, 72

Economics of photovoltaic power systems,
  171–196, 198
  central power systems, 188–196
  intermediate level systems, 186–188
  residential systems, 179–188
  small-scale systems, 54, 171
Edge-defined film-fed growth of Si,
  130–132, 166, 199
Efficiency
  effect of base doping density, 97
    grain properties, 121–123
    material purity, 126, 132
    minority carrier properties, 118–120
    semiconductor thickness, 120–121, 134
    series and shunt resistance, 116–118
    sunlight concentration, 156–159
    temperature, 159–162
  effect on concentrator costs, 193–194
  limit calculations, 110
  of arrays, 21, 40, 56
  of CdS cells, 111, 113, 116, 143–147
  of GaAs cells, 111–112, 123, 136, 148,
  150, 155, 162–164, 199
  of InP cells, 123
  of organic cells, 152
  of Schottky barrier cells, 112–113, 148
  of Si cells, 20, 110, 112–113, 122–123,
  132, 134, 136–138, 141, 156, 159,
  164–170, 199
  of tandem cells, 151
Electrolysis, 12, 42, 48–49
Electron affinity, 100, 104
Encapsulation, 22–23, 32, 57, 163–164
  adhesive, 23
Energy payback time, 153
Energy storage, 4–5, 11, 41–51
  large systems, 171–196
  sizing of, 54, 56, 180, 190–196
  small systems, 9, 21, 52, 54, 56, 171

Epitaxy, 68
 in CdS cells, 113
 in GaAs cells, 112, 150, 162, 164
 in Si cells, 103, 108, 120, 132, 134, 136,
  167
Equivalent circuit, solar cell, 69, 105

Fed-back power, 41–42, 180, 183, 186
Figure of merit, F_M, 180–181, 187
Fill factor, 70–71, 107–108
 factors affecting, 104–107, 113, 117, 119,
  155, 157–158, 160
 in CdS cells, 108, 147
 in GaAs cells, 107
 in Si cells, 108, 134, 158
Fixed charge rate, 175
Flat plate arrays, 22–27, 174–179, 199
 see also Arrays
Fuel cells, 44–45, 49
Fuel escalation rate, 180, 185
Flywheel storage, 45, 49–50

Gallium arsenide cells
 absorption in, 68–69, 84, 87, 120
 concentrator cells, 155, 157–159,
  161–164, 178–179, 196, 199
 dark currents, 94, 98
 efficiency, 110, 112, 114
 fill factor, 107
 flat plate cells, 116, 122, 136, 148–150
 hybrid systems, 196
 J_{sc}, 77, 79, 91
 Schottky barrier cells, 87–88, 92, 100,
  104, 108, 113, 119, 122
 spectral response, 81, 83, 88
 V_{oc}, 99, 104
GaAs/AlAs cells, 80, 84, 92, 107, 111, 164
 energy band diagram of, 80
 thin film, 150
GaAs/Au cells, 87–88, 92, 108, 119
GaAs/I/Au cells, 104, 108, 113
GaAs/Ga$_{1-x}$Al$_x$As cells, 33, 35, 136, 146,
  162–164
 efficiency, 111–112, 155, 164
 fill factor, 107
 J_{sc}, 91–92
 spectral response, 83–84
 V_{oc}, 99, 104

GaAlSb, 148
GaAgSe$_2$, 148
GaCuSe$_2$, 148
GaP/Si cells, 81–83, 99
GaP.$_{22}$As.$_{78}$/Au cells, 108
Generation, electron-hole pairs, 72, 122
Generation mix, 4, 11, 44, 65, 173
Glassy metal oxides, 108, 138, 140
 see also Indium tin oxide, Tin oxide
Graded bandgap semiconductors, 83–84, 92,
  108, 111, 120
Grain boundaries, 105, 115, 117, 133
 diffusion down, 122–123, 138, 147
 effect on efficiency, 121–123
 recombination at, 85, 115, 122
Granular Si, 122, 132–133, 199
Grating cells, 138, 140

Heat exchanger method of Si growth,
  127–128
Heat sinks, 32–33, 36, 161, 176
Heavy doping effects, 95, 97, 103, 110, 160
Heterojunctions, 67–68, 107, 160
 dark current, 103–105
 efficiency, 110–113
 fill factor, 107–108
 J_{sc}, 79–85, 91–92
 thin film, 116, 121, 141–148, 150
 V_{oc}, 98–100, 103–104
Homojunctions, 67–68, 122, 124–125, 156,
  159
 efficiency, 109–110, 112, 122
 fill factor, 105, 107
 J_{sc}, 74–79, 91
 shunt resistance, 117
 thin film, 134, 136–137, 146
 V_{oc}, 93–98, 103–104

Indium phosphide
 absorption in, 69, 120
 Schottky barrier cells, 122
InP/CdS cells, 92, 104, 108, 112–114, 116,
  142, 145–147
Indium tin oxide, 91, 103, 113, 141, 148
 see also Glassy metal oxides
Injection
 high level, 157–158, 160, 166
 low level , 74, 156, 158

Insolation, 9, 12–21, 52
  diffuse, 17–18
  direct, 14–18
  on inclined plate, 16, 55
  variability factor, 51, 54, 172
Institutional factors, 41, 62–65, 173–174
Interconnection, solar cells, 2, 23–26,
    32–33, 36–37, 52, 114, 124, 144–145
Interconnection, solar/nonsolar plant, 4,
    9–11, 37–43, 64–65, 172–196, 198
Interface states, 84, 99, 100, 102, 104, 141,
    147, 150
Inversion layer cells, 138, 140
Inverters, 10, 38–39, 186
Ion implantation, 68, 138–139, 153

Kerf losses, 126, 153

Lateral pulling of Si, 129
Learning curve, 5, 63
Lifetime, photovoltaic system components,
    22, 171, 173, 197
Load factor, 4, 44, 64, 172
Load leveling, 41, 46, 187

Markets, photovoltaics, 5, 12, 23, 37, 144,
    147, 174, 197
Maximum power point, 11, 23, 39, 70, 105,
    107
Metallurgical grade Si, 124–126, 134
Minority carrier
  diffusion length
    effect of high doping, 77–79
    effect of temperature, 159–161
    in BSF cells, 97–98
    in GaAs, 92, 112, 158–159, 162
    in polycrystalline material, 115,
      118–119, 122
    in Si, 89, 98, 159, 167–169
  lifetime
    effect of high doping, 95
      temperature, 159
    in CdS, 92
    in CdTe, 148
    in GaAs, 98
    in large area cells, 118–119
    in Si, 89, 110
  Schottky barrier cells, 102, 119, 140

Mismatch, lattice, 81, 84, 92, 99, 104, 107,
    113, 145
Mismatch, solar cell, 21, 40, 151
MIS solar cells, 138, 140
  dark current, 93, 101–102, 117, 119
  efficiency, 112–113
  fill factor, 108
  in Cu_2O, 150
  in GaAs, 104, 108, 113, 148, 150
  in Si, 93, 108, 112
  J_{sc}, 90
  reflection losses, 90–91
  V_{oc}, 101–104
Module, solar cell *See Array*
  costs, 187–188
Multijunction solar cells, 167–170

Nuclear power, 1, 198, 200

Ohmic contacts, 72, 134
  recombination velocity of, 120
  to CdTe, 146
  Cu_2O, 151
  GaAs, 148, 150, 163
  InP, 147
  organics, 152
  Si, 165
Open circuit voltage, 2, 70–71, 108, 113,
    156
  effect of bandgap, 71, 109, 151
    conductivity modulation, 158
    dark current, 93–103, 119, 178
    diode exponential factor, 105, 107
    resistivity, 95–98, 140
    temperature, 155, 159–161
    thickness, 120, 134
  practical values, 103–104, 134, 136
  thin film cells, 134, 147
Organic semiconductors, 116, 152
Ownership, photovoltaic power plant, 41,
    64–65, 172, 179, 182, 187, 198

Packing factor
  modules, 175–178, 186
  solar cells, 20–21, 26–27, 126, 133, 144
Peak shaving, 42, 187
Peeled film technology, 134–135, 164

Photocurrent suppression, 90, 101, 108, 115

Photolithography, 117, 137, 165

Photovoltage saturation, 156–157

Photovoltaic power systems
central, 3, 7, 27, 37–38, 43, 72, 188–196
intermediate, 4, 7, 27, 37–38, 43, 72, 186–188
residential, 3–4, 7, 27, 37–38, 43, 172, 179–186
satellite, 12, 56–62, 64, 171
small-scale, 5–7, 9, 12, 23, 42–43, 51–56, 63, 144, 171

Pollution, 5, 22, 60–61, 171, 173

Polycrystalline semiconductors
CdS, 92, 141–147
Cu$_2$O, 149–150
current leakage, 105, 117
GaAs, 148–150
grain boundary effects, 115, 118, 121–123, 138
Si, 124, 133, 136

Power conditioning, 10, 21, 37, 42, 171, 194

Power tracking, maximum point, 10, 39, 42

Production facility, solar cells, 139, 143–144

Production rate, solar cells, 116, 124, 126
CdS, 143–144
Si, 23, 124, 126, 133, 138
U.S. goals, 3, 5

Pumped hydro storage, 45, 47–48

Pyrolitic spraying, 121, 143–144

Recombination, photocarrier
at back contact, 97, 120
in heterojunctions, 84–85, 148
in interfacial layer, 108, 148

Reflection losses, 85, 90–91, 110, 113, 140

Resistance, load, 41, 70

Rheotaxy, 136

Ribbon to ribbon Si growth, 131–132

Schottky barrier solar cells, 67–68, 138, 140
effect of temperature, 160
efficiency, 110, 112, 119
fill factor, 105, 108

J$_{sc}$, 85–92
V$_{oc}$, 100–104
with amorphous Si, 137
organics, 152
polycrystalline material, 119, 122, 148, 150

Screen printing, 117, 137

Secondary impurities in Si, 125, 133

Selenium, 66

Semiconductor grade Si, 124–126, 153

Series resistance, 69–71
effect on fill factor, 104–108
in CdS cells, 92
concentrator cells, 155–158, 163–165
GaAs cells, 92, 150, 155, 157, 163
large area cells, 24, 116–118
organic cells, 152
Si cells, 91, 156–158, 164–165, 167, 169

Short circuit photocurrent, 2, 71–74, 108
effect of
bandgap, 109
base resistivity, 97
diode exponential factor, 105
minority carrier properties, 118–120
sunlight concentration, 156
temperature, 159, 161
thickness, 120–121
in heterojunctions, 79–85
homojunctions, 74–79
large area cells, 118, 134
Schottky barriers, 85–91, 119
practical values, 91–92

Shunt resistance, 69–70
effect on efficiency, 116–117
fill factor, 104–105

Silicon cells, 67, 114–115, 147, 199
absorption in, 68–69
amorphous cells, 136–137, 199
array costs, 176–180, 187
concentrator cells, 32, 156–161, 164–170, 199
dark currents, 94–97
effect of temperature, 159–161
efficiency, 20, 110–114, 119
encapsulation, 23
energy payback time, 153
fill factor, 105–108, 157–158
grain boundary effects, 122–123
granular cells, 122, 132–133, 199

J_{sc}, 76, 91
large area cells, 117, 119–120, 137–141, 199
minority carrier properties, 79, 118, 120
multijunction cells, 167–170
preparation methods
  film, 133–137
  sheet, 129–133
  wafer, 126–129
  purity, 124–126
Schottky barrier cells, 87, 89, 91, 101, 108, 113, 117, 119
spectral response, 81, 83
surface states, 100
V_{oc}, 98–99, 103–104
Si/Au cells, 87–89, 119
Si/GaP cells, 81–83, 99
Si/ITO cells. *See* Indium tin oxide
Si/SiO$_x$/Al cells, 93, 108, 112
Si/SnO$_2$ cells. *See* Tin oxide
Solar grade silicon, 124–126, 128, 137
Spectral current density, 76, 87–88
Spin-on doping, 137–138
Stability, solar cells, 22, 141, 143–147, 150, 163
Stepanov growth of Si, 131
Superconducting magnet energy storage, 45, 50–51
Surface recombination velocity, 74
  at grain boundaries, 115, 122
  in BSF cells, 97, 120
  in CdTe, 146
  in GaAs, 79, 91, 111
  in Si, 79, 120, 122, 169

Tandem cells, 151–152
Temperature, 4, 154
  effect on efficiency, 159–163
Textured surface cells, 91, 112, 120, 137
Thermal energy, 137, 159, 196, 199
Thickness, semiconductor, 120–122, 133, 143, 150, 152, 166
Thin film solar cells, 115, 118
  CdS, 116, 120–121, 141–146
  CdTe, 120, 145–146, 148
  Cu$_2$O, 149–151
  CuInS$_2$, 146–147
  CuInSe$_2$, 146–147
  GaAs, 120, 136, 148–150
  InP, 120, 146–147
  Organics, 152
  Si, 112, 120, 133–137, 199
Tin oxide, 91–92, 103, 113, 121, 141, 143
  See also Glassy metal oxides

Utilities, 4, 7, 11, 21, 38, 41, 43, 46–47, 49–50, 64–65, 172–200
Utility charge rate, 180, 188

Wafer slicing, 126, 128

Zinc-based solar cell materials
  Zn$_3$P$_2$, 148
  ZnS, 92
  ZnSe, 81–82, 99, 148
  ZnSiAs$_2$, 148
  ZnTe, 148